Ethics and the Digital Transformation of Human Work

Peter G. Kirchschlaeger

Ethics and the Digital Transformation of Human Work

The Society-, Entrepreneurship-, Research-Time Model (SERT)

palgrave
macmillan

Peter G. Kirchschlaeger
Institute of Social Ethics ISE
University of Lucerne
Luzern, Switzerland

ISBN 978-3-031-98703-8 ISBN 978-3-031-98704-5 (eBook)
https://doi.org/10.1007/978-3-031-98704-5

Cover illustration: Pattern © Melisa Hasan

This Palgrave Macmillan imprint is published by the registered company Springer Nature Switzerland AG.
The registered company address is: Gewerbestrasse 11, 6330 Cham, Switzerland

To our daughter Mia Esther

Foreword

Writing a book resembles a journey that one travels mostly alone and explores curiously new horizons. In the case of traveling, the moments that create probably the most beautiful memories are the encounters and interpersonal interactions with other people. In the case of writing a book, the conversations with other humans matter and represent a wonderful source of amazing inspiration and meaningful ideas. It was definitely like that with this book, and I'm deeply grateful for it.

My special thanks go to the faculty members and researchers of the Philosophy Department and of the Divinity School at Yale University for the in-depth discussions that I was able to have with them as a visiting fellow at Yale University during the academic years 2015–2017, for their interest in my research, and for their contribution to the conceptualization of this study.

For important contributions during the development of this study provided through guiding questions, essential commentaries, and inspiring discussions, I am grateful to my colleagues at the Faculty of Theology of the University of Lucerne, the team of the Institute of Social Ethics ISE at the Faculty of Theology of the University of Lucerne Sonia Arfaoui, Antonia Bilic, Melina Faeh, Matteo Frey, Sander de Haan, Ursina Huber, Dr. Alexandra Kaiser-Duliba, Dr. Juerg Kuehnis, Annamaria Muff-Ricci, Andrea Murer, Joshua Samuel Schibli, Nicolas Stilter, and Dr. Evelyne Tauchnitz and its former team members I had the honor to work with Dr. Aaron Butler, Silvia Choinowski, Monika De Col, Estefania Cuero, Dr. Ana Laura Edelhoff, Selina Frankhauser, Sonja Freitag, Dr. Noemi Honegger, Kiki Kuenzler, Shania Kuhn, Nataliia Manko, Ramatu Musa,

Dr. Karin Nordstroem, Birgit Rath, Dr. Anna Maria Riedl, Kaja Schmid, Claudia Schmutz, Carmen Staub, Chantal Studer, Oleksandra Trajkovska, Dr. Ernst von Kimakovitz, Yan Wagner, Juliette Wyler, and the participants of the Lucerne Graduate School in Ethics LGSE at the ISE (including Jean-Marie Kuzituka Did'ho, Elizaveta Ebner, Prof. Dr. Thomas Gremsl, Sara Ilić, Ilja Kirianoff, Mojalefa Koloko, Dr. Laurence Lerch, Dr. Darius Meier, Fr. Leonard Nnamdi, Jan Thomas Otte, Nina Stern, Adrienne Hochuli Stillhard, Stefanie Uhl, and Matej Vereš) with whom I was able to lead topical discussions. I'm especially grateful for the support with the formal preparation of the manuscript by Antonia Bilic, Ursina Huber, Sander de Haan, Annamaria Muff-Ricci, and Joshua Samuel Schibli.

I am also very grateful for the constructive discourse with my colleagues at the University of Lucerne; the ETH Zurich; the Global Ethic Institute at the University of Tuebingen; the Swiss Federal Ethics Committee on Non-Human Biotechnology (ECNH); the Ethics-Advisory Board Smart Lucerne of the City of Lucerne; the working group Mobility 4.0 of the Swiss Federal Roads Office (FEDRO) at the Swiss Federal Department of the Environment, Transport, Energy and Communications (DETEC); the Focus Group (a small committee of experts in the field of Artificial Intelligence) serving the Advisory Council on Digital Transformation of the Swiss Federal Council; the Board of Trustees of the Switzerland Foundation of the European Academy of Sciences and Arts; the Scientific Advisory Board of the European Forum Alpbach; the Advisory Board of the Think Tank "Dezentrum," a Think Tank on New Technologies of Trust Square (Zurich's Blockchain hub); the think tank "RedBit"; the Council for a 21st Century Progressive Economy; the members of the faculties of the Annual Lucerne Summer University: Ethics in a Global Context LSUE under the patronage of UNESCO 2018–2024, and the guest speakers of the "EthicsImpulses" that have been held thus far at the Institute of Social Ethics ISE of the Department of Theology of the University of Lucerne, especially Professor Wilfried Stadler (WU Vienna University of Economics and Business and University of Salzburg).

The research-oriented environment of the Faculty of Theology of the University of Lucerne, where I was able to carry out this research, significantly contributed to the creation of this study. I'm grateful that I was able to present the thesis and results of my research during the development phase to many panels as well as at various conferences and events and discuss my thoughts with several dialogue partners, such as the colleagues and participants at, among others, the Annual Meeting of the "Innsbrucker

Kreis von Moraltheolog*innen und Sozialethiker*innen" 2025, the SFTL—SwissFinTechLadies Gala at the World Economic Forum WEF in Davos 2025, the VNW-Managementforum Hamburg (Germany) 2025, the 64th Aerztekongress Davos 2025, the HUMAN Project Berlin (Germany) 2025, the PhiloThiK in the Theatre of the Kornhaus Baden (Switzerland) 2025, the Congress My Hope 2025 in Goerlitz (Austria), the International Lucerne Ethics of Human Rights-Symposia LES 2023 and 2024, the International Symposium "Global Regulation for 'Artificial Intelligence'" 2024, the Hanseatic Compliance Day of the Chamber of Commerce Hamburg (Germany) 2024, the m4music festival Zurich 2024, the Linecheck Music Meeting and Festival Milano (Italy) 2024, the Conference of Language Center of UZH and ETH Zurich 2024: VOICE(S), the NCCR Automation Annual Retreat of ETH Zurich 2024, the Internet Governance Forum IGF Austria 2024, the International Conference of the Genshagen Foundation Berlin (Germany) 2024, the Annual Conference of the Swiss Association of Research Ethics Committees Swissethics 2024, the Assembly of Delegates of UNICEF Switzerland 2024, the "MA RWP Lecture" of the Center for Religion, Economy and Politics (CREP) 2024, the TecDay & TecNight at the High School Kantonsschule Wohlen of the Swiss Academy of Engineering Sciences SATW 2024, the Religion-Forum of the High Schools of the Canton Argovia 2024, the Annual Conference of the HSG Alumni Seniors Club 2024, the Annual Lucerne Summer University: Ethics in a Global Context LSUE under the patronage of UNESCO 2018–2024, the International Conference: Right to a Future, Rights of a Child 2024, the Global Peter Drucker Forum Vienna (Austria) 2021 and 2024, the Lichtenberg Dialogue Berlin (Germany) 2024, the Athens Democracy Forum, Athens (Greece) 2024, organized by the Democracy & Culture Foundation in association with the New York Times, the Frankfurter Buchmesse (Germany) 2024, the Haus Gutenberg-Conference (Liechtenstein) 2024, the Swiss AI Impact Forum 2024, the Impact Days Vienna (Austria) 2023 and 2024, the Congress of Creative Entrepreneurship, Schliersee (Germany) 2024, the Edufest 2024, the World Summit on the Information Society (WSIS)+20 Forum High-Level Event 2024, the Internet Governance Forum 2023 at Kyoto (Japan), the International Research Workshops on "Ethics of Digital Transformation" 2023, the UNCTAD eWeek 2023: Shaping the future of the digital economy, the Integrity Europe Conference 2023, the SDG Digital at United Nations Headquarters in New York (USA) 2023, the Science Summit at the UN General Assembly in New York (USA) 2023, the High Level Side Event of the

53rd Session of the UN Human Rights Council 2023, the WSIS Forum 2023, the World Z-inspection® Conference Venice (Italy) 2023, the AI+X Summit of ETH Zurich 2022, the Annual Conference of the Global Justice Program at Yale University 2022, the Internet Governance Forum 2022 at Addis Ababa (Ethiopia), the Fritz-Erler-Forum Stuttgart (Germany) 2022, the Internet Governance Forum 2021 at Katowice (Poland), the Finance and Economy Forum 2021, the "IDEEN Lab: Artificial Intelligence and Ethics" of the Oesterreichische Forschungs foerderungsgesellschaft FFG 2019, the NZZ X.Days 2018 in Interlaken, the Swisscom Dialog Arena 2018, the European Forum Alpbach 2018, the International Ethics in Action-Conference "The Future of Work" in the Vatican 2018 convened and chaired by Professor Jeffrey Sachs (Director of the Earth Institute, Columbia University; Special Advisor to the UN Secretary-General António Guterres on the objectives of sustainable development; Director of the UN Sustainable Development Solutions Network), the Globart Academy 2017, and the Annual Seminar for the Swiss Financial Market Supervisory Authority FINMA 2017, as well as the colleagues and students of institutions such as the University of Lucerne, the ETH Zurich, the University of Tuebingen (Germany), the University of St Gallen, the University of Zurich, the Chung-Ang University Seoul (South Korea), the Sophia University Tokyo (Japan), the SCMHRD, an AACSB accredited institute at Pune (India), the University of Vale do Rio dos Sinos, São Leopoldo (Brazil), the University of Parma (Italy), the University of Roma Tre (Italy), the University of Salzburg (Austria), the University of Graz (Austria), the Institute of Science and Technology in Gwangju (South Korea), the Seoul National University of Education (South Korea), the University of Fribourg, the University of Applied Sciences Lucerne, and the University of Oxford (UK).

I would like to thank Amy Invernizzi of Palgrave Macmillan in particular for including this study in their program. When handing in the manuscript for print, Antony Sami and Soniya Sankar were very helpful, for which I'm thankful.

My wife Miriam and our daughters Sara and Mia, my entire family, and my friends followed the development of this research with their love and friendship, with encouraging interest and great support. All my thanks go to my "small world."

Lucerne, Switzerland — Peter G. Kirchschlaeger
March 31, 2025

Reviews of *Ethics and the Digital Transformation of Human Work*

"Peter G. Kirchschlaeger delivers an insightful exploration into the ethical dimensions of the digital transformation reshaping our workplaces. Through engaging analysis, he examines the disruptive impact of 'artificial intelligence' on human labour and offers a thought-provoking framework for society on how to embrace the opportunities of innovation while addressing its ethical challenges."

—Christine Lagarde, *President of the European Central Bank*

"BIG TECH is inventing the AI-saturated future that serves their interests and ambitions for profit and social control. Sadly, most of us depend on their explanations and justifications explaining AI Developments that are conducted in secret, outside of democratic governance, and intrinsically opaque. Our societies rely on the Tech giants' accounts of the inevitable consequences of their work, and their shocking cluelessness as to its implications for the lives of humans, our societies, and our politics. Peter G. Kirchschlaeger boldly chooses the road less traveled, with striking erudition and profound commitment to the human, he offers a deeply creative framework for a human future that co exists with new forms of machine intelligence that serve humanity rather than control us. Peter G. Kirchschlaeger assembles an intricate framework of social structure and human activity based on the moral principle of human dignity, the psychology of human fulfillment, and the politics of society as the enduring guarantor of human rights and democratic self-governance. This is a future owned and operated by humanity, not by Tech overlords, and everyone on earth needs more thinking like this. The future is ours to make, not the property of private corporations to impose. We need deep systemic creative conceptions built on the fundamental primacy of humanity and the justice, beauty, care, flourishing and fulfillment that only we can produce together. Peter G. Kirchschlaeger's work is an antidote to the feelings of hopelessness and helplessness that so often overtake us. It asks us to join the fight for a human future."

—Shoshana Zuboff, Author *The Age of Surveillance Capitalism; Charles Edwar Wilson Professor Emeritus, Harvard Business School, USA*

"For most of us the words 'Artificial Intelligence' conjure up a maze of conflicting predictions and assertions. Will it help humanity? Will it overwhelm us? Who can guide us through this maze and help us to understand the phenomenon, its potential, its power and its problems? Peter G. Kirchschlaeger's scholarly and prescient work is precisely the trustworthy and accessible resource we need to navigate this rapidly changing digital world, to make sense of it and to ensure it is used ethically in the benign service of creation."

—Professor Dr. Dr. h.c. mult. Mary Patricia McAleese, *former President of Ireland; Chancellor of the Trinity College Dublin, Ireland*

"In the midst of the maddening AI noise, a structured work of reflection built on the art of asking the right questions."

—Silvio Napoli, *Executive Chairman of the Board of Directors of Schindler Group*

"Data-based systems pose challenges for the future of work that differ from earlier technological revolutions, as this important book makes crystal clear. It is not just about job numbers, but about the nature of work and the ethical principles that guide it and that form society. A must read for researchers and policy makers in this area."

—Professor Dr. Katharina Pistor, *Edwin B. Parker Professor of Comparative Law; Director, Center on Global Legal Transformation, Columbia Law School, Columbia University, USA*

"If you want to read a book about the impact of digital transformation and AI on human labor, read this book! It is a brilliant and sharp scientific analysis combined with an inspiring and concrete proposal to create an ethically sound future for everyone and the planet: SERT."

—Anat Bar-Gera, *Tech Investor, Board Member, Serial Entrepreneur*

"Lucidly and with great intellectual depth, this book addresses one of the major societal questions: How will AI affect the nature of work and human dignity? Its main thesis is that unlike in previous technological disruptions, this time it will be different."

—Professor Dr. Helmut Bölcskei, *Professor of Mathematical Information Science, Department of Electrical Engineering and Department of Mathematics, ETH Zurich*

"This book discusses the ethical dilemmas posed by the digital transformation of human work. Arguing that Data-Based Systems (DS) will likely displace more jobs than they create, it proposes the innovative SERT model: a system of decoupled income and societal contribution designed to ensure a just and dignified future for all. A timely and important analysis with concrete solutions."

—Professor Dr. Thomas Pogge, *Leitner Professor of Philosophy and International Affairs, Yale University, USA*

"In an era characterized by careless enthusiasm for data-based systems—over-heroically labelled 'artificial intelligence' and shaped by economic greed—the work (Ethics and the Digital Transformation of Human Work. The Society-, Entrepreneurship-,

Research-Time Model SERT) is a refreshingly necessary demystification of the greatest voodoo since the beginning of digital evolution. Lucid and pragmatic, it provides answers to questions that too few people have even asked so far."

—Nina George, *Prize-Winning and International Bestselling Author, President of Honor of the European Writers' Council*

"To read this book is an absolute must for everyone who is interested and/or involved in the development of social welfare. Peter G. Kirchschlaeger captures not only possible systemic far-reaching impacts of AI- and Digital Transformation on society at large, but also how the purpose of life and the future identity of human-beings would need to be addressed by all of us. It also gives hope for those who cannot yet see the positive side of our technological developments."

—Stefan Bachofen, *Chairman of bbv Software Services Group AG*

"An outstanding book! Everyone who is interested in the future of human work—and that's probably almost all of us—should read this groundbreaking and scientifically rigorous study on the impact of 'AI' on human work from an ethical perspective. It does not only analyze ethical downsides and upsides of this disruption but offers with SERT also a concrete, inclusive, and sustainable model as a way forward that is convincing in the global 'north' and 'south'."

—Professor Dr. Kutoma Wakunuma, *Professor of Research and Teaching in Information Systems at the School of Computer Science and Informatics of the Faculty of Computing, Engineering and Media of De Montfort University, UK and Zambia*

"I highly recommend this book, which warns of advanced AI systems displacing jobs at an unprecedented scale, challenging the assumption that technology leads to new job creation. Peter G. Kirchschlaeger's SERT model proposes systemic safeguards to uphold human dignity and economic stability, providing a crucial framework for leaders to align AI with societal values before irreversible disruptions occur."

—Sungjin (James) Kim, *Senior Applied Scientist at Selling Partner Services (SPS), Amazon*

"'Ethics and the Digital Transformation of Human Work' by Prof. Peter G. Kirchschlaeger is a timely and deeply thoughtful contribution to one of the most pressing challenges of our era. With clarity and moral urgency, Peter G. Kirchschlaeger examines the disruptive potential of digital transformation and so-called Artificial Intelligence on human labor. A must-read for policymakers, technologists, and ethicists alike."

—Professor Dr. Benjamin F. Grewe, *Professor of Neuronal Learning and Intelligent Systems; Institute of Neuroinformatics UZH/ETH Zurich/Dept. of Electrical Engineering and Information Technology, ETH Zurich*

"Innovative and unconventional, this book enriches the currently evolving debate about the future of work with fresh ideas and original insights into what the work should represent for the human species in the era of expanding digitalization. A rare example of academic work, where ethics meets digital transformation and turns into practical suggestions on how to improve our lives as citizens and workers. A must read to those who look for thought-provoking approaches to the future of work."

—Professor Dr. Migle Laukyte, *Associate Professor in Artificial Intelligence and Law, Pompeu Fabra University in Barcelona, Spain; Member of the European Group on Ethics in Science and New Technologies (EGE)*

"Ethics and the Digital Transformation of Human Work is a powerful read that challenges the assumption that technological progress always leads to job growth. It explores how Data-Based Systems (DS) reshape work, sometimes creating jobs but often displacing them or altering skill demands. The book highlights the 'black-box problem' of algorithmic opacity, emphasizing the need for transparency and accountability in data driven economies. Grounded in human rights ethics, it advocates for dignity, equality, and autonomy in the face of digital transformation. It also explores solutions like unconditional basic income and the Society-, Entrepreneurship-, Research-Time (SERT) model to address job displacement. A key takeaway is its call for global governance through an UN-backed International Data-Based Systems Agency (IDA) and a framework of 30 guiding principles. For anyone working at the intersection of AI, economics, and ethics, this book is a must read. It doesn't just offer critique; it lays out solutions, making it an invaluable resource for those of us striving to shape a more just and responsible digital future."

—Shamira Ahmed, *Executive Director of the Data Economy Policy Hub (DepHUB), South Africa*

"Peter G. Kirchschlaeger is peering into the future with human lucidity. While his analysis recognizes the inevitability of the digital paradigm shift, he provides guidance on shaping this evolution, pathways that preserve core human rights, most notably the right to live a fulfilling life with dignity. My takeaway… humankind will not capitulate!"

—Professor Dr. Philip D. Jaffé, *University of Geneva, Member of the UN Committee on the Rights of the Child*

"How can we ensure that the coming digital transformation of human labor enhances human flourishing, even as it leads to a radical reduction of paid professional work? Peter G. Kirchschlaeger moves beyond the tired alternatives of apocalyptic catastrophe and technophilic utopia to develop a detailed assessment and a concrete proposal for a systemic ethical response. An essential resource."

—Professor Dr. Jennifer Herdt, *Gilbert L. Stark Professor of Christian Ethics; Professor of Religious Studies; Senior Associate Dean for Academic Affairs, Yale University Divinity School, Yale University, USA*

"This book offers a timely and insightful contribution to the critical discourse on digital transformation and data-based systems (DS). In this book, Peter G. Kirchschlaeger examines the ethical dimensions of rapidly evolving technologies and their impact on both professional and private lives. Beyond analyzing the challenges to human work, the book explores innovative models and advocates for a Human Rights-Based approach to DS and the establishment of an International Data-Based Systems Agency (IDA) at the UN. For anyone seeking a comprehensive and ethically grounded understanding of the societal implications of AI and potential pathways towards a humane and sustainable future, this book provides valuable guidance and thought-provoking perspectives."

—Professor Dr. Damian Okaibedi Eke, *Faculty of Science, University of Nottingham, UK, and Zambia*

"In a time of profound, technology-driven changes in the world of work, Peter G. Kirchschlaeger's book is a vital contribution for policymakers, researchers, and professionals navigating an uncertain future. It serves both as a reminder of the pressing ethical challenges we face today and as a much-needed proposal for how we might address them."

—Dr. Yuri Lima, *Future of Work Research Coordinator, Future LAB, Federal University of Rio de Janeiro (UFRJ), Brazil*

"When machines do our work, what will be left for us? In his book, Peter G. Kirchschlaeger addresses the central question of how the remaining work, welfare and social participation must be distributed to enable everyone to live a life of human dignity and proposes ethical ways of addressing the significant challenges associated with the digital transformation of human work."

—Professor Dr. Sabine T. Koeszegi, *Head of the Institute of Management Science, Professor of Labor Science and Organization, Academic Director of the Executive MBA Program IDEa, TU Wien, Austria; Chair of the UNESCO Advisory Board on the Ethics of Artificial Intelligence in Austria; Member of the AI Advisory Board of the Austrian Government*

"Peter G. Kirchschlaeger calls for a human-centered digital transformation that prioritizes human dignity and rights over technological determinism. Technology must remain a means to an end—designed to serve people, their values, and the common good. His Society-, Entrepreneurship-, Research-Time (SERT) model offers an alternative vision for work and societal participation beyond traditional employment. Indeed, ethical reflection is not an optional luxury but the foundation for ensuring that digital innovation fosters justice, solidarity, and a sustainable, human-rights-based future."

—Professor Dr. Thomas Beschorner, *Professor for Business Ethics and Director of the Institute for Business Ethics, University of St Gallen, Switzerland*

"While everyone hopes to improve the human condition through digital transformation, there are diverse and even conflicting views on how to achieve this. In this context, SERT, proposed by Peter G. Kirchschlaeger, will serve as a valuable theoretical platform for social discussions on real-world issues."

—Professor Dr. Insok Ko, *Professor of Philosophy & Dean of the College of Humanities, Inha University, Korea; former President of the Korean Society for the Philosophy of Science*

"This is essential reading for anybody interested in a profound analysis of the digital transformation of human labor and a compelling, practices-oriented vision of a future where technology serves humanity and not the other way around."

—Professor Dr. Catrin Misselhorn, *Professor of Philosophy, University of Goettingen, Germany*

"The book of Peter G. Kirchschlaeger examines the ethical implications of digital transformation and artificial intelligence on human labor, highlighting systemic risks such as job displacement and economic inequality and their impact on human dignity, and identity. The Society-, Entrepreneurship-, Research-Time (SERT) model proposed offers a visionary framework for addressing these implications in our era of rapid technological change. The book offers some sobering thoughts for those of us developing automation and digitalization solutions about the potential consequences of our work."

—Professor Dr. John Lygeros, *Professor for Computation and Control and Head of the Automatic Control Laboratory, Department of Information Technology and Electrical Engineering, ETH Zurich*

"This visionary project aims at no less than making digital transformation of human work a means for transforming economy into an instrument of equal dignity for all human beings—that is an inspiring challenge."

—Professor em. Dr. Philippe Mastronardi, *Professor Emeritus of Law, University of St Gallen*

"The book *Ethics and the Digital Transformation of Human Work* is a serious and well-documented call for democratic control of so-called Artificial Intelligence. Based on a sober and sobering assessment of AI (it's incapacity for moral action and accountability, the fact that its purpose is to exclude humans from work rather than facilitate it, the market pressures that accelerate its implementation), Peter G. Kirchschlaeger makes a serious and detailed call for Human Rights-Based Data-Based Systems under international regulation, in order to guarantee that such systems be safe, transparent, and at the service of human rights, which are the moral grounding of the work."

—Professor Dr. Germàn Bula, *Professor of Philosophy, Universidad Pedagogica Nacional, Bogota, Colombia*

"Peter G. Kirchschlaeger skillfully analyses the ethical impact of 'data-based systems' on human dignity, freedom, well-being, and capacity to meaningfully contribute to society, while at the same time not shying away from proposing concrete and actionable visions for new political and economic initiatives to support an ethical transition to an increasing digitalized society. Most importantly, he avoids the temptation of speculating about hypothetical futures and then describing the challenges that humanity will face in those futures; he rather directly tackles the central philosophical, ethical, and political question of the present time: how should we as a society think and act today to fulfil our responsibility to a future where technology supports deep human, social, and environmental needs and not just the political, technological or economic agenda of current technology developers?"

—Professor Dr. Filippo Santoni de Sio, *Professor of Ethics of Technology, TU Eindhoven, The Netherlands*

Contents

CHAPTER 1

Introduction

Abstract This first chapter is dedicated to the presentation of the aim and method of the book. The aim of this book is to provide an ethical analysis of the digital transformation of human work and of the impact of so-called "Artificial Intelligence (AI)" on human labor. Digital transformation and the use of so-called "AI"—more adequately defined as "Data-Based Systems (DS)"—are already changing human lives today. They open new horizons for innovating business models and economic systems as well as new forms of work. This dynamic will create new jobs. At the same time, the disruptive impact of digital transformation and DS on human labor deserves close consideration.

Keywords Digital transformation • Artificial intelligence • Data-based systems (DS) • Human labor • Ethical analysis • Innovative business models • New form of work • Disruption

1.1 Aim and Method of This Book

Digital transformation and the use of so-called "Artificial Intelligence" has already changed the everyday professional and private lives of humans. "Digital technologies and data have dramatically changed the way people live and work, how and in which markets and firms operate, and the ways

P. G. Kirchschlaeger, *Ethics and the Digital Transformation of Human Work*, https://doi.org/10.1007/978-3-031-98704-5_1

in which governments interact with citizens" (Organisation for Economic Co-operation and Development [OECD], 2024, p. 78). Economic and social transformation occurs now, not just in the future (European Commission and Servoz, 2019). Digital transformation and "AI" open new horizons for innovating business models and economic systems as well as new forms of work. This boost will contribute to solving social and ecological problems. This dynamic will create new jobs. "We expect new roles will arise but we are unsure how long they will last, because these too, in due course, may be taken on by machines" (Susskind & Susskind, 2015, p. 405). The only certainties encompass, first, the technology-based change of human work and, second, the uncertainty of what this transformation of labor will contain (AI Action Summit, 2025, p. 214).

> Artificial intelligence (AI) has the capacities to ease our existence as it supports the growth of productivity and innovation. At the same time, its disruptive impact on labor markets and settings of work as well as on further education and training – specifically skills demand and skills supply – deserves close consideration. Therefore, action is needed for just and timely upskilling and reskilling to adapt skills to changed requirements and framework conditions. (Larsen & Kipper, 2022, p. 13)

At the same time, technology-based transformation of business processes leads to challenges for workers, employers, and policymakers (Korinek, 2024; McAfee, 2024), and disruptions (Freeman, 2015) as well as at the same time relief for humans and creates new space for other tasks. The question arises whether or not digital transformation and the use of "AI" will lead to a massive reduction in paid professional tasks for humans (Willcocks, 2020), leading to "technological unemployment" as it was coined by John Maynard Keynes: "unemployment due to our discovery of means of economizing the use of labor outrunning the pace at which we can find new uses for labor" (Keynes, 1930, para. 1).

Beyond that, this impact of technological progress on human labor will proceed in manifold ways in different regions, countries, and contexts of the globe.

> The uneven adoption of Artificial Intelligence (AI) is a critical issue that goes beyond economic growth. It impacts global equity, fairness and the social contract that is at the heart of social justice. Disparities in access to robust infrastructure, advanced technology, quality education and training

> are deepening existing inequalities. As the global economy increasingly shifts towards AI-driven production and innovation, less developed countries risk being left further behind, exacerbating economic and social divides. Without targeted and concerted efforts to bridge this digital divide, AI's potential to foster sustainable development and alleviate poverty will remain unrealized, leaving significant portions of the global population disadvantaged in the rapidly evolving technological landscape. (United Nations [UN] and International Labor Organization, 2024, p. 3)

These issues are obviously of ethical relevance. The ethical analysis of this question will include a comparison with earlier technology-based eras of change in human history. Beyond that, the digital transformation of human work and the impact of "AI" on human labor need to be ethically assessed. Finally, concrete ways are to be identified fostering the ethical opportunities and mastering or avoiding the ethical risks of the effects of technology-based changes to human work.

> AI does not happen to us; choices made by people determine its future. How general-purpose AI is developed and by whom, which problems it is designed to solve, whether we will be able to reap its full economic potential, who benefits from it, and the types of risks we expose ourselves to – the answers to these and many other questions depend on the choices that societies and governments make today and in the future to shape the development of general-purpose AI. Since the impact of general-purpose AI on many aspects of our lives is likely to be profound, and since progress might continue to be rapid, there is an urgent need to work towards international agreement and to put resources into understanding and addressing the risks of this technology. Constructive scientific and public discussion will be essential for societies and policymakers to make the right choices. (AI Action Summit, 2025, pp. 214–215)

The aim of this book is to address these and related questions in order to contribute to ethical guidance for the design and the realization of digital transformation of human work and the impact of "AI" on human labor. It is necessary because often in studies of ethics, the reflection, analysis, and discussion from an ethical standpoint of the question of the impact of digital transformation and "AI" on human work is neglected (see, e.g., Grimm et al., 2024). The methodological approach will consist of the methods of theoretical ethics.

1.2 Data-Based Systems (DS) Rather Than "Artificial Intelligence"

1.2.1 Intelligence?

The ethical analysis of the digital transformation and the impact of so-called "Artificial Intelligence" on human work starts with a critical examination of terms from an ethical standpoint. So-called "artificial intelligence" can be defined as striving by technical means to imitate or fulfill cognitive functions of human thought. During an ethical critique of so-called "AI," it becomes clear that it does not comprise the sum of human knowledge, nor is it objective, fair, and neutral. It is not trained by or does not represent reality, facts, or scientific evidence, but is trained by and represents the online past. It is only based on certain data.

In the case of so-called "generative AI" (e.g., ChatGPT), the image of the cow chewing the cud applies, because generative "AI" actually does nothing other than absorb as data what people have already thought, written, developed, painted, photographed, or created, then mixes it up once and then spits it out again generously. Strictly speaking, nothing new is created. Impressively, what is spat out here proves to be syntactically correct and semantically coherent. However, despite all the appreciation, it is also important to remain critical in the sense that one should really ask: What is this spit-out actually about? And here it should be noted that what people have already thought, written, developed, painted, photographed, or created is being reproduced in a remixed form. Generative "AI" is virtually stuck in the past, it remains a prisoner of its own data basis, which is only composed of what is already known.

Beyond that, it does not matter at all to generative "AI" whether it is telling the truth or not. Generative "AI" does not even care whether what it spits out is right or wrong. It's all about filling information channels with content, regardless of whether this content is true or false. The syntax and semantics must be right. The truth criterion applies to people, even if people do not always adhere to it. When people communicate with each other, they generally assume that they are telling each other the truth, even if this is not always the case. Interpersonal communication only makes sense if adherence to the truth criterion is fundamentally assumed. Otherwise, one could save oneself the time for the conversation. With people, people generally assume that there is a desire to tell the truth, and that the truth criterion plays a role.

As data becomes scarce, the data used for the training in the course of the further development of generative "AI" includes more and more the data that so-called "generative AI" generates itself. As a result, so-called "AI" is getting dumber and dumber.

Looking at the above-mentioned definition of so-called "AI" from an ethical perspective, the term "artificial" is not questioned because this technology is created by humans.

A first criticism arises regarding "intelligence" because intelligence does not just consist of the solution of a cognitive task but also in the way it is pursued (Misselhorn, 2018). A second criticism focuses on the nature of so-called "AI" imitating human intelligence (Dreyfus, 1972; Dreyfus & Dreyfus, 1986). It is to be assumed that so-called "AI" can at best imitate human intelligence but can never become the same. A third criticism highlights that so-called "AI" is limited to certain areas of intelligence (e.g., certain cognitive capacities). Among others, in the domain of emotional and social intelligence, machines are only able to simulate emotions, personal interaction, and relationships but lack authenticity. For instance, a health care robot can be trained to cry when the patient is crying, but no one would argue that the robot feels real emotions and cries due to them. The robot does not care about it. On the contrary, one could train the exact same robot to slap the patient's face when the patient is crying, and the robot would perform this function in the same perfect way. Again, the robot does not care about it. The lack of authenticity of robots in health care is problematic for respecting the dignity of all humans (Turkle, 2007). As it is relevant to the respect of human dignity, authenticity must be part of the equation in data-based health care and in the use of "care robots" (Manzeschke, 2019). Machines cannot reach emotional and social intelligence—neither today nor tomorrow—as the expectable further increase of compute of machines in the future can indeed improve the simulation of emotions by machines but does not create emotional and social intelligence.

1.2.2 Moral Capability?

Beyond that area of human intelligence, in the domain of moral capability, one cannot ascribe machines with moral capability because they are presupposed to follow ethical patterns and rules given by humans. Technologies are primarily made for their suitability and may set rules as a self-learning system, for example, to increase their efficiency, but these rules do not contain any ethical quality. For example, a self-driving car could set the rules for itself, but it is not aware of the ethical quality of these rules. It could give itself the rule to get from A to B as fast as possible including harming humans and nature, to optimally fulfill the task of reaching B in the shortest time possible, without being able to recognize ethical rules for itself, which would allow the machine to perceive the illegitimacy of its rules and actions. A human driver instead possesses the potential to recognize for themself binding ethical rules, which empower them to see that harming humans and nature might be more efficient but illegitimate. Machines lack this autonomy. Autonomy encompasses recognizing and setting ethical norms for oneself and basing one's own actions on them. Humans can set ethical rules for a self-driving car, whether good or bad. Machines are not able to recognize by themselves for themselves ethical rules because they fail on the principle of generalizability. This principle has its roots in Immanuel Kant's universalization principle that ethical rules can only be ethical rules if one wants them to be universal law (Kant, 1974). Based on this, the fulfillment of the principle of generalizability presupposes presenting rational and plausible arguments—"good reasons". "Good reasons" means that it must be conceivable that all humans, given their effective freedom and autonomy as well as their full equality, would agree upon these reasons—within a model of thought and not within a real worldwide referendum—on ethical grounds (Kirchschlaeger, 2023). While humans are able to recognize by themselves ethical rules for themselves by fulfilling the principle of generalizability and perceive them for themselves as binding, machines cannot. On this basis, humans can know that they do something ethically right or wrong while machines cannot identify the ethical quality of their accomplishment of tasks. *They don't even care* if they fulfill a task in a legitimate or illegitimate manner, or whether it is a legitimate or illegitimate task. The potential that technologies possess in relation to ethical decisions and actions is nowhere close to moral capability because machines lack not only autonomy but also vulnerability, conscience, freedom, and responsibility, which are all essential for human morality (Kirchschlaeger, 2021).

Therefore, technologies cannot perform as moral subjects or moral agents (Kirchschlaeger, 2021), but humans carry exclusively the ethical responsibility for machines. Humans must lay down ethical principles as well as ethical and legal norms, set a framework, goals, and limits for digital transformation and machines, and define the use of machines in addition to examining, analyzing, evaluating, and assessing technology-based innovation from an ethical perspective. Then, machines can follow ethical rules given to them by humans.

It is plausible that machines would be even more strict in obeying these rules because they just do what one tells them to do. For example, a self-driving car will respect the traffic rules and will probably cause substantially fewer accidents because it does not drive drunk, nor testosterone-induced, tired, distracted by texting, lost in thought....

1.2.3 The Rule Transcending Uniqueness of the Concrete

There are two limitations to this "ethical" performance of machines. The first one we already know: Machines are not aware that they are doing something legitimate or illegitimate. *They don't even care* if their actions are ethically positive or negative. In other words, one could program or train machines to do something ethically unacceptable and they would implement it with the same rigor—without any access to the ethical quality of the task or the way of accomplishing the task.

Second, through the increasing complexity of everyday reality, ethical principles or norms can diverge or collide. If this occurs, humans are challenged to find ethical insights into the ethical assessment of a concrete encounter with concrete persons in a concrete situation. Sometimes ethics must go beyond principles, norms, and rules to be sensitive to *the rule-transcending uniqueness of the concrete* (Kirchschlaeger, 2023). This accounts for the truth that in a concrete encounter with concrete people in a concrete situation, rules can reach their limit because the concrete, in its uniqueness outranks the rule.

> The general, concrete ethical, the positive legal and many other norms that are generally applicable, although indispensable, are not sufficient to guarantee the basic humanity, in the face of diversity (...). It is inevitable that we have to cross norms in certain situations in order to act humanely, but this does not mean that we deny the need for norms in general or refute that they are generally applicable. (Virt, 2007, pp. 42–43, translated by the author)

These ethical considerations in a more differentiated and better manner would be expecting too much of machines due to their lack of moral capability. The following example can illustrate this: Imagine a situation in the Nazi era in which we are hiding a Jewish family from the Nazis in our home. Suddenly the Nazis knock on the door and ask us if we are giving shelter to a Jewish family. If we follow the truth, we send the Jewish family to certain death. If we strive to save the lives of the Jewish family, we must lie to the Nazis. So, what is the ethically right thing to do in this concrete situation, in this concrete encounter with concrete people? In this case, saving the lives of the Jewish family should be given higher priority than the truthfulness requirement. Therefore, lying to the Nazis in this concrete situation, in this concrete encounter with concrete people is the ethically correct thing to do. Ethics is characterized by its sensitivity to *the rule-transcending uniqueness of the concrete.*

Concerning the rule-transcending uniqueness of the concrete, firstly, it is crucial that this rule-breaking is not done to enrich oneself or to serve a self-interest, a preference, a wish, or a desire, but to achieve a higher ethical good in this concrete situation, in this concrete encounter with concrete people. Secondly, ethical and legal norms and their validity are not called into question. Thirdly, they are reinforced by this striving for the ethically right and good. Fourthly, the sensitivity of ethics to the rule-transcending uniqueness of the concrete ensures that ethical norms serve people and not vice versa. Fifthly, this does not call into question the meaningfulness and raison d'être of ethical norms in general in the sense of a call for anarchy or anomy.

The necessity for ethics to be sensitive to the rule-transcending uniqueness of the concrete shows that transferring ethics completely to mathematics or programming and training of machines becomes difficult or even impossible. Due to the lack of moral capability, it is impossible for machines to practice and respect the rule-transcending uniqueness of the concrete which is an essential characteristic of ethics.

1.2.4 Interactionist Model

At the same time, one must also take into account that technologies (Borgmann, 1992), robots (Gunkel, 2023), and especially "smart" technologies are influencing—for example, by nudging as discussed by Mathis and Tor (2016)—at least individual lives and maybe even the ethical dimension of individual lives (Guthrie, 2013). "By helping to shape human actions and experiences, technologies also participate in our ways of doing ethics" (Verbeek, 2011, p. 1). While acknowledging the effect technology has on humans as subjects of ethics, technologies—in contrast to the idea of a "nonhumanist ethics of technology" (among others Verbeek, 2011, pp. 21–40)—remain objects of ethics and human subjects of ethics. Why? Humans can decide on an ethical basis or orientation towards an ethical frame of reference in the technological decisions they make: whether to create a technology or not, how and if they design, develop, produce, disseminate, and use a technology or not, if they abolish a technology or not, if—to a certain extent—they allow a technology to influence their lives or not, and how they assess a technology from an ethical perspective. All this is up to humans—the subject of ethics—not up to technologies, the objects of ethics. This cannot be adequately captured by either the personification of machines (Verbeek, 2011) or by the construction of a middle-ground between personification and reification (Gunkel, 2023).

This ethical analysis provokes at first sight the criticism that it takes ethical reasoning back to a point where it was before and got liberated from, namely, perceiving the relation between humans and technologies as a conflict between the two. This is not the intention. Rather, it tries to reconcile Michel Foucault's "critique" (Foucault, 1984, pp. 32–50) with the subjectivity of humans distinguished from the objectivity of technologies by acknowledging "the fundamental intertwinement" (Verbeek, 2015, p. 215) of humans and technologies implying "that the frameworks from which we criticize technologies are always mediated by these technologies themselves" (Verbeek, 2015, p. 215), and by re empowering the critical characteristics of ethics, embracing also the possibility of questioning fundamentally technology-based innovation and technologies as such. Both ethics from inside and from outside technologies are ethically necessary, and humans are able to strive for both—on the conceptual basis of understanding the relationship between technologies and ethics as interaction (Kirchschlaeger, 2021; Kirchschlaeger, 2024b).

The ethical foundation for this reconciliation embraces, on the one hand, the self-critical acceptance of the boundaries of humans and human reason that makes it impossible for humans to liberate themselves completely from the intertwinement of humans and technologies.

On the other hand, it trusts in the subjectivity of humans allowing them rationally to distance themselves enough from technologies in order to analyze, discuss, evaluate, assess, and judge the object "technologies" from an ethical perspective.

Without this latter aspect, humans would run the risk of being trapped again in "immaturity", accepting authorities and following their orders as criticized by Immanuel Kant—in present times, obeying immaturely the authorities of technological progress, innovation, and economic imperatives.

Without this latter aspect, humans would be reduced to immature and powerless addressees, users, and consumers of powerful technologies (Weizenbaum, 1976).

Without this latter aspect, technologies would represent an absolute and fatalistically determined reality not influenceable by humans, expressed representatively, for example, in the following way: "Technological development will continue, and human existence will change with it" (Verbeek, 2015, p. 214).

Of course, humans are exposed to and intertwined with technologies while, for example, holding a gun:

> You are different with a gun in hand; the gun is different with you holding it. You are another subject because you hold the gun; the gun is another object because it has entered into a relationship with you. The gun is no longer the gun-in-the-armory or the gun-in-the-drawer or the gun-in-the-pocket, but the gun-in-your-hand, aimed at someone who is screaming. What is true of the subject, of the gunman, is as true of the object, of the gun that is held. A good citizen becomes a criminal, a bad guy becomes a worse guy; a silent gun becomes a fired gun, a new gun becomes a used gun, a sporting gun becomes a weapon. The twin mistake of the materialists and the sociologists is to start with essences, those of subjects *or* those of objects. That starting point renders impossible our measurement of the mediating role of techniques. Neither subject nor object (nor their goals) is fixed. (Latour, 1994, p. 33)

This mediation needs to be considered because it is capable of covering the way humans interact with the world.

> On the one hand, the concept of mediation helps to show that technologies actively shape the character of human-world relations. Human contact with reality is always mediated, and technologies offer one possible form of mediation. On the other hand, it means that any particular mediation can only arise within specific contexts of use and interpretation. Technologies do not control processes of mediation all by themselves, for the forms of mediation are always context-dependent. (Verbeek, 2005, p. 11)

This mediation offers an alternative to an "instrumentalist" understanding of technologies (neutral means for humans to reach their aims) and to a "substantivist" understanding of technologies (determining and controlling influence on the individual, the society, and culture) (Verbeek,

2005, pp. 1–12). Exceeding these approaches, the *interactionist understanding of the relation between technologies and ethics* (Kirchschlaeger, 2021) thus coins this rapport as dynamic—knowing different grades of intensity concerning how humans, technologies, and the world interact with each other, and embracing different kinds of compositions, how much each of them contributes to creating an ever-new reality from a human perspective.

For example, understanding the shovel as a mediator between humans and the ground like the artist Tomi Ungerer (Ungerer, 2019, p. 18), the *interactionist model* comprises the possibilities of mediation, of an instrumentalist narrowing of technologies, of a substantivist elevation of technologies, of ideological human self-image defined by technologies to the point of self-abandonment in favor of technologies under the illusion of technologies as an end in itself, and techno-critical reluctance of technologies as well as their dynamic combination in manifold compositions. The interactionist approach situates ethics in dialogue with and under influence of this dynamic and of these possibilities. It understands ethics with the ability to distance itself to a certain extent—enough for an ethical analysis, discussion, evaluation, assessment, and judgment of technologies. This interactionist understanding of technologies and ethics trusts in humans remaining, at the end of the day, able as subjects to use the shovel or to put the gun as an object down and to reflect upon this tangible shovel or gun in this specific context as technology; to discuss in a democratic process the production, the accessibility, the sale, the use, the non-use, or the destruction of shovels or guns; to form a political opinion and make a political decision about it; and to reflect upon shovels or guns in general as technologies from an ethical standpoint, as well as technologies in general from an ethical perspective. This interactionist understanding of technologies and ethics trusts humans to be capable of creating, or of not producing, of using or not using, or of destroying technologies under ethical guidance. It empowers humans to reflect upon technologies and to explore the lack of moral capability of machines.

1.2.5 Ethical Relevance

The neglect of moral capability for technologies does not only not mean that machines cannot follow ethical rules. As discussed above, they are able to obey and to apply ethical rules rigorously. The neglect of moral capability does not mean either that their actions do not matter ethically.

Of course they do. A self-driving car obeying the ethical rule that it should not harm humans makes an ethical difference compared to a self-driving car rushing from A to B as fast as possible no matter what—including harming humans. Therefore, contributions by technologies to human lives are ethically relevant and can be ethically evaluated (Verbeek, 2011, pp. 1–20).

This ethical relevance of "human making and using" (Mitcham, 2014, p. 11), of "tools and their deliberate use" (Bloch, 1959, p. 731), as the foundation of civilization, though, needs to be distinguished from possessing ethical subjectivity or from moral agency. For example, the above-mentioned self-driving car is ethically relevant because it can save human lives, or it is able to do harm to or kill humans. But a self-driving car as a material object—even though it represents a highly sophisticated technology—is not an ethical subject or a moral agent because, among other things, it cannot be held accountable for harm or killing. Humans behind this technology remain the ethical subjects or the moral agents and are accountable. They make decisions about the interactivity (response to stimulus by change of state), "autonomy" (ability to change without stimulus), and adaptability (ability to change the "transition rules" by which the state is changed) of technologies; they create, design, develop, produce, disseminate, abolish, use, or not use technologies. Behind the supposedly "political bridges" (Winner, 1980), behind the supposedly "missing masses of morality" (Latour, 1992), behind the supposedly "artificial moral agents" (Floridi & Sanders, 2004; Hanson, 2014; Introna, 2014), behind the supposed "artefacts with morality" (Verbeek, 2011, pp. 41–65), behind the supposedly "moral agents and mediated subjects" (Verbeek, 2011, pp. 66–89), and behind the supposed "morality in design" (Verbeek, 2011, pp. 90–119) there are humans as ethical subjects and moral agents responsible for these technologies.

In order to illustrate this point with other examples: humans can decide whether or not to build bridges in a way that is disadvantageous to humans – manifesting their political convictions in a lasting way. Humans can decide, for example, whether or not to build speed bumps in order to protect human lives. Humans can decide to create, design, develop, build, produce, disseminate, use or not use a technology which can perform a "morally qualifiable action." "An action is said to be morally qualifiable if and only if it can cause moral good or evil" (Floridi & Sanders, 2004, p. 361). Humans can decide if a technology should perform this action, and if a technology should be able to decide by itself if it should perform this action. Just because a technology is ethically relevant as it can cause ethically positive and ethically negative effects does not mean it possess moral agency as there is, *first*, a difference between performance and moral agency. *Second*, it is an ethical decision by humans if a technology should be entrusted with such decisions. *Third*, even in the case of a technology being designed by humans in a way that it makes "autonomous" decisions about ethically relevant questions, a technology does not make these decisions with a knowledge, perception, or awareness of the ethical quality of these decisions.

Humans can make ethical decisions about mediation by technology, its intensity, and its extent. Humans can decide to create, design, develop, build, produce, disseminate, use or not use a technology which may be able to mediate, and humans can decide if a technology should mediate or not. Humans can be aware of mediation by technologies, and humans can reflect self-critically upon mediation by technologies. Humans are not exposed powerlessly and heteronomously to it.

1.2.6 Data-Based Systems (DS)

Based on these considerations from an ethical standpoint acknowledging the general potentials and deficiencies of so called "artificial intelligence" as technologies and the specific potentials and deficiencies of so-called "artificial intelligence," the term "artificial intelligence" raises ethically justifiable doubts because these technologies do not represent "intelligence". The term *"data-based systems (DS)"* (Kirchschlaeger, 2021; Kirchschlaeger, 2022) would be more appropriate than "artificial intelligence" because this term describes what actually constitutes "artificial intelligence": generation, collection, and evaluation of data; data-based perception (sensory, linguistic); data-based predictions; data-based decisions. In addition, the term "data-based systems (DS)" allows for highlighting the main strengths and weaknesses of the present technological achievements in this field. The mastery of an enormous quantity of data depicts the key asset of these technologies—of *data-based systems (DS)*.

Pointing to its core characteristic, namely being based on data and relying exclusively on data in all its processes, its own development, and its actions—more precisely, its reactions to data—lifts the veil of the inappropriate attribution of the myth of "intelligence," covering substantial ethical problems and challenges of data-based systems (DS). This allows more accuracy, adequacy, and precision in the critical reflection on DS. For instance, the untrace ability, unpredictability, and inexplicability of the algorithmic processes resulting in data-based evaluation, data-based predictions and data-based decisions ("black-box-problem") (Bathaee, 2018; Castelvecchi, 2016; Knight, 2017a, 2017b), its wide vulnerability to systemic errors, its deep exposure for confusing causality with correlation (e.g., high consumption of ice cream by children in a summer month and high number of children car accidents due to more mobility during vacation in the same summer month correlate but there is not any causal relationship between the two statistics, meaning ice cream consumption does not cause car accidents) (Iversen & Gergen, 1997), and its high probability of biased and discriminatory data leading to biased and discriminatory data-based evaluations, predictions, and decisions embrace its major disadvantages (Coeckelbergh, 2020; UN Educational, Scientific and Cultural Organization [UNESCO], 2024). "Algorithms are opinions embedded in codes. They are not objective" (Demuth, 2018). They are not neutral. They are not fair. They serve specific goals and purposes (e.g., economic, political, ...). They are not trained by or do not represent the reality, facts, or scientific evidence, but are trained by and represent the online past. They are only based on certain data.

Finally, this terminological sharpening of course does not exclude the possibility of relying on and learning from the existing research and discourse on so-called "AI" (including, e.g., "knowledge-based systems") and its technological and normative dimensions.

1.2.7 Exclusive Human Responsibility for DS

The previous considerations show that DS lack moral capability and are neither "moral subjects" nor "moral agents" but objects. If DS do not possess moral capability, moral subjectivity, or moral agency, they cannot bear responsibility. For example, if a self-driving vehicle should cause an accident, one cannot hold accountable the self-driving vehicle because it is not able to be a subject of responsibility. The plausibility of this argument can be illustrated with the imagination of the following practice: It does not make sense and does not contribute to justice, restitution, compensation, or to making amends if the self-driving car were punished, for example, by destroying it or by unplugging it from power supply for two weeks. Or one cannot blame DS for taking away jobs from humans.

Based on these elaborations, humans bear the exclusive responsibility for DS. This responsibility embraces ethical decision-making as well.

Moreover, the definition of ethical and legal principles and norms as well as a framework for digital transformation and DS also belong in the hands of humans. This exclusive responsibility of humans for digital transformation and DS becomes even more important due to the following characteristic of digital transformation:

> The special feature of digitalization is not so much that hybrid machines such as robots and automated systems are playing an increasingly important role in everyday life, but rather the change in accountability that goes hand in hand with digital processes. Digital processes are essentially characterized by the duplication and multiplication of independent causal relationships, which lead to an increased dependence on automated decision-making processes. The exponential increase in computing power and data volumes exceeds the cognitive processing capacity of the users of digital systems and their regulatory control of these systems. (Heidbrink, 2024, p. 63, translated by the author)

The exclusive responsibility of humans for digital transformation and DS actually starts with the question whether a technology should be created or not. Digital transformation and DS do not simply fall from the sky or are not predetermined by nature. Humans carry the responsibility to examine, analyze, and evaluate technology-based innovations from an ethical point of view and not to realize everything that is technically feasible. It is up to humans to decide whether, what, and how digital transformation and DS are designed, developed, produced, deployed, and used or not used for ethical reasons. This also means that humans are responsible for the contexts and working conditions of the extraction of the natural resources which are necessary for digital transformation and DS as well as for the entire supply chain including all the production sites. The entire value-added chain must be ethically acceptable; thus, among others, environmental protection standards and human rights must be respected, protected, implemented, and realized. The question to use or not use a technology also includes the responsibility to use digital transformation and DS in a targeted manner where they serve humanity and the planet, and to develop DS, for example, in those areas of human intelligence where DS already massively outperform humans—such as handling large amounts of data, data storage, computing power and logical deduction. DS can substantially support human thinking by storing facts and perceiving complexity.

Part of this exclusive human responsibility for digital transformation and DS also consists in taking into account the complexity of ethics (Kirchschlaeger, 2024a), which includes—besides the above-mentioned rule-transcending uniqueness of the concrete—the "dual use" problem and the ambivalence (Kirchschlaeger, 2023). The "dual use" problem (Rath et al., 2014) is based on the potential of the same DS to be used for either a legitimate or an illegitimate purpose (European Commission, 2018; Miller, 2018; Sevini et al., 2018). It originates from the characteristic of DS as multiuse tools. This dual use problem can be illustrated with the example of a drone: Drones can be used to deliver essential food to people in disaster zones where other people cannot reach them. But the same drones can be used for killing: the so-called "killer drones." Of course, the ethical evaluation of such machines will be diametrically opposed in those two situations. This dual use problem shows the limits of the general evaluation of drones. For a sophisticated ethical analysis, a general assessment of drones needs to be complemented by considering the uses of drones.

The dual use problem is recognized in its relevance by the "regulatory conceptual framework" entitled "Dual Use Research of Concern (DURC)" (World Health Organization, n.d.) used by national and international organizations such as the World Health Organization (WHO) or the US National Institute of Health (NIH).

The description of the dual use problem as "deviation of intent" (DiEuliis & Giordano, 2018, p. 239) proves to be too reductionist, because the possibility that a research and innovation process may also begin with an ethically negative objective also needs to be considered. As stated above, it cannot be automatically assumed that innovation processes always aim for an ethically positive purpose.

At the same time, it can be stated that the dual use problem contains less than a dilemma, as the US National Research Council tries to grasp it conceptually (Imperiale & Casadevall, 2015), because it does not necessarily have to have a dilemma structure and thus, an ethical analysis is certainly able to differentiate the ethically positive from the ethically negative. Looking at digital transformation and DS from an ethical perspective, it means that not only the technology as such needs to be examined, but also its concrete uses and applications.

Ambivalence though describes the possibility that DS may be used for an ethically good purpose but can at the same time include or lead to something ethically bad. Differing from Hans Jonas' "ambivalence of effects" (Jonas, 1985, pp. 42–43), ambivalence embraces the actions, that is to say, the cause as well as the effects (Kavanagh, 2019; Scholz et al., 2018). For example, a DS-learning assistant fosters the learning performance of a student which is ethically positive because it helps the student. At the same time, while using the DS-learning assistant, the student becomes a product themself because their data is stolen and sold to a third party.

The following ethical analysis of the impact of DS on human work in this book will also include recent developments like generative DS (so-called "generative AI"). Generative DS will not be addressed separately for one technological, one economic, and three ethical reasons:

From a technological perspective, generative DS are not unique in their impact on human work as they may continue the ongoing amplification of impact by DS on human labor based on the further development of reached domains and skills by DS.

From an economic perspective, generative DS specifically impact more the conditions of human work rather than the automation of human work:

> The immediate effect of the most recent wave of generative AI will be neither growth in automation nor the emergence of new industries, but the decrease in difficulty of existing content-creating jobs. [...] AI has considerably expanded the potential scope of automation. But thinking of the most recent wave of generative AI – which, it must be noted, is merely a subfield of AI – as an automation technology is, in its current form, a mistake. For one thing, generative AI requires humans to make a prompt and then select (as well as mostly edit) the desired output. This prompting and selection is where much of the actual creativity resides. It is more apt to think about generative AI as analogous to Uber and its impact on taxi services. With the advent of GPS technology, knowing the name of each street in New York City was no longer a valuable skill. So, when Uber rolled out across the United States, average drivers with little knowledge of the cities in which they operated took full advantage. The result was not fewer jobs, but more intense competition, which reduced the incomes of incumbent drivers. In joint work with Lund University's Thor Berger, our lab found that drivers' hourly earnings fell by 10% when Uber entered a given city. (Frey & Osborne, 2024)

From an ethical perspective, generative DS, first, do not change the relationships of responsibility. In other words, the responsibility for DS—even for generative DS—remains in the hands of humans.

Second, the substitution of humans on the labor market by DS is still occurring. It is even growing due to the amplification of domains covered by DS, fueled by generative DS as well. Even if the latter is not or not entirely the case, the negative impact on working conditions for humans would also be of ethical relevance as well as it would emphasize the necessity of continuing to offer humans the possibility to experience the functions so far fulfilled by paid professional tasks within a working-context respecting human dignity.

Third, DS cause in this way more "unpaid work" and therefore also contribute to the massive reduction of paid professional tasks. "The amount of unpaid 'work' is rising relative to the amount of paid 'labor', as more people work away from fixed workplaces and outside fixed working hours, or with jobs that pay only when 'on-task', ignoring travel time (in the case of, for example, care workers), waiting time (drivers), security check time (Amazon warehouse workers), and so on" (Standing, 2018, p. 115).

For the ethical analysis of digital transformation of human work and the impact of the use of DS on human labor in this book, the elaborations so far should allow a more precise understanding of the technologies one is dealing with as well as a clear idea about the exclusive human responsibility for digital transformation and DS. Humans decide what is done and what is not. Humans decide what is happening and what is not. Humans decide if and how digital transformation and DS should affect themselves as humans and human work.

In order to live up to their exclusive responsibility for digital transformation of human work and the impact of DS on human labor, the question needs to be clarified whether digital transformation of human work including the use of DS will create or destroy paid jobs for humans.

References

AI Action Summit. (2025). *International AI safety report. The international scientific report on the safety of advanced AI.* Retrieved March 31, 2025, from https://www.gov.uk/government/publications/international-ai-safety-report-2025

Bathaee, Y. (2018). The artificial intelligence black box and the failure of intent and causation. *Harvard Journal of Law & Technology, 31*(2), 889–938.

Bloch, E. (1959). *Das Prinzip Hoffnung: Kapitel 1–32* (Werkausgabe, Band 5). Suhrkamp.

Borgmann, A. (1992). The moral significance of material culture. In A. Feenberg & A. Hannay (Eds.), *Technology and the politics of knowledge* (pp. 85–93). Indiana University Press. https://doi.org/10.1080/00201749208602295

Castelvecchi, D. (2016). Can we open the black box of AI? *Nature, 538*(7623), 20–23. https://doi.org/10.1038/538020a

Coeckelbergh, M. (2020). *AI ethics. The MIT Press essential knowledge*. MIT Press. https://doi.org/10.7551/mitpress/12549.001.0001

Demuth, Y. (2018, April 26). Die unheimliche Macht der Algorithmen. *Beobachter*. Retrieved March 31, 2025, from https://www.beobachter.ch/digital/multimedia/big-data-die-unheimliche-macht-der-algorithmen

DiEuliis, D., & Giordano, J. (2018). Gene editing using CRISPR/Cas9: Implications for dual-use and biosecurity. *Protein & Cell, 9*(3), 239–240. https://doi.org/10.1007/s13238-017-0493-4

Dreyfus, H. L. (1972). *What computers can't do: The limits of artificial intelligence*. MIT Press.

Dreyfus, H. L., & Dreyfus, S. E. (1986). *Mind over machine: The power of human intuition and expertise in the era of the computer*. Free Press. https://doi.org/10.1109/MEX.1987.4307079

European Commission. (2018). *Knowledge for policy. Dual-use technologies*. Retrieved March 31, 2025, from https://ec.europa.eu/knowledge4policy/foresight/topic/changing-security-paradigm/artificial-intelligence-quantum-cryptography_en

European Commission: European Political Strategy Centre & Servoz, M. (2019). *AI, the future of work? Work of the future! On how artificial intelligence, robotics and automation are transforming jobs and the economy in Europe*. Publications Office. https://data.europa.eu/doi/10.2872/913422

Floridi, L., & Sanders, J. W. (2004). On the morality of artificial agents. *Minds and Machines, 14*(3), 349–379. https://doi.org/10.1023/B:MIND.0000035461.63578.9d

Foucault, M. (1984). What is enlightenment? In P. Rabinow (Ed.), *The Foucault reader* (pp. 32–50). Pantheon Books.

Freeman, R. (2015). Who owns the robots rules the world. *IZA World of Labor*. https://doi.org/10.15185/izawol.5

Frey, C. B., & Osborne, M. (2024). Generative AI and the future of work: A reappraisal. *Brown Journal of World Affairs, 30*(1), 1–17. https://bjwa.brown.edu/30-1/generative-ai-and-the-future-of-work-a-reappraisal/

Grimm, P., Trost, K. E., & Zoellner, O. (Eds.). (2024). *Digitale Ethik*. Nomos. https://doi.org/10.5771/9783748942399

Gunkel, D. J. (2023). *Person, thing, robot. A moral and legal ontology for the 21st century and beyond*. MIT Press. https://doi.org/10.1007/s10676-023-09731-9

Guthrie, C. F. (2013). Smart technology and the ethical life. *Ethics & Behavior, 23*(4), 321–337. https://doi.org/10.1080/10508422.2013.787359

Hanson, F. A. (2014). Which came first, the doer or the deed? In P. Kroes & P.-P. Verbeek (Eds.), The moral status of technical artefacts (Vol. 17, pp. 55–73). Springer. https://doi.org/10.1007/978-94-007-7914-3

Heidbrink, L. (2024). *Nichtverantwortlichkeit. Zur Deresponsibilisierung der Gesellschaft*. Velbrueck Wissenschaft. https://doi.org/10.5771/9783748951599

Imperiale, M. J., & Casadevall, A. (2015). A new synthesis for dual use research of concern. *PLoS Medicine, 12*(4), e1001813. https://doi.org/10.1371/journal.pmed.1001813

Introna, L. D. (2014). Towards a post-human intra-actional account of socio-material agency (and morality). In P. Kroes & P.-P. Verbeek (Eds.), The moral status of technical artefacts (Vol. 17, pp. 31–53). Springer. https://doi.org/10.1007/978-94-007-7914-3_3

Iversen, G. R., & Gergen, M. (1997). *Statistics: The conceptual approach. Springer undergraduate textbooks in statistics*. Springer. https://doi.org/10.1007/978-1-4612-2244-6

Jonas, H. (1985). *Technik, Medizin und Ethik. Zur Praxis des Prinzips der Verantwortung*. Suhrkamp.

Kant, I. (1974). Grundlegung zur Metaphysik der Sitten. In W. Weischedel (Ed.), *Werkausgabe* (Vol. 7, 7th ed.). Suhrkamp. (Original work published 1785).

Kavanagh, C. (2019). *New tech, new threats, and new governance challenges: An opportunity to craft smarter responses?* Carnegie Endowment for International Peace. Retrieved March 31, 2025, from https://carnegie-productionassets.s3.amazonaws.com/static/files/WP_Camino_Kavanagh_New_Tech_New_Threats1.pdf

Keynes, J. M. (1930). Economic possibilities for our grandchildren. In J. M. Keynes (Ed.), *Essays in persuasion* (pp. 358–373). Macmillan & Co. (Original work published 1930, collection published 1931). https://gutenberg.ca/ebooks/keynes-essaysinpersuasion/keynes-essaysinpersuasion-00-h.html

Kirchschlaeger, P. G. (2021). Digital Transformation and Ethics. Ethical Considerations on the Robotization and Automation of Society and the Economy and the Use of Artificial Intelligence. Nomos.

Kirchschlaeger, P. G. (2022). Ethische KI? Datenbasierte Systeme (DS) mit Ethik. *HMD Praxis der Wirtschaftsinformatik, 59*(2), 482–494. https://doi.org/10.1365/s40702-022-00843-2

Kirchschlaeger, P. G. (2023). Ethical Decision-Making. Nomos. https://doi.org/10.5771/9783748918684

Kirchschlaeger, P. G. (2024a). Artificial intelligence and the complexity of ethics. *Asian Horizons, 18*(3), 375–389. https://dvkjournals.in/index.php/ah/article/view/4590/3752

Kirchschlaeger, P. G. (2024b). *In an era of digital disruptions, ethics can't be an afterthought – Part 1 and part 2*. Business and Human Rights Journal Blog. Retrieved March 31, 2025, from https://www.cambridge.org/core/blog/2024/04/11/in-an-era-of-digital-disruptions-ethics-cant-be-an-afterthought/

Knight, W. (2017a, April 11). The dark secret at the heart of AI. *MIT Technology Review*. Retrieved March 31, 2025, from https://www.technologyreview.com/s/604087/the-dark-secret-at-the-heart-of-ai/

Knight, W. (2017b, April 13). The financial world wants to open AI's black boxes. *MIT Technology Review*. Retrieved March 31, 2025, from https://www.technologyreview.com/s/604122/the-financial-world-wants-to-open-ais-black-boxes/

Korinek, A. (2024). *Economic policy challenges for the age of AI* (NBER Working Paper No. 32980). National Bureau of Economic Research. https://doi.org/10.3386/w32980

Larsen, C., & Kipper, J. (2022). Introduction. In C. Larsen, J. Kipper, A. Schmid, & M. Ricceri (Eds.), *The relevance of artificial intelligence in the digital and green transformation of regional and local labour markets across Europe: Perspectives on employment, training, placement, and social inclusion* (pp. 13–26). Hampp. https://doi.org/10.5771/9783957104113-95

Latour, B. (1992). Where are the missing masses? The sociology of a few mundane artifacts. In W. E. Bijker & J. Law (Eds.), *Shaping technology / building society* (pp. 151–180). MIT Press.

Latour, B. (1994). On technical mediation: Philosophy, sociology, genealogy. *Common Knowledge, 3*(2), 29–64.

Manzeschke, A. (2019). Roboter in der Pflege: Von Menschen, Maschinen und anderen hilfreichen Wesen. *Ethik Journal, 5*(1), 1–11. Retrieved March 31, 2025, from https://www.ethikjournal.de/fileadmin/user_upload/ethikjournal/Texte_Ausgabe_2019_1/Manzeschke_1.Nov_FINAL.pdf

Mathis, K., & Tor, A. (Eds.). (2016). *Nudging – Possibilities, limitations and applications in European law and economics*. Springer. https://doi.org/10.1007/978-3-319-29562-6

McAfee, A. (2024). *Generally faster: The economic impact of generative AI*. Retrieved March 31, 2025, from https://policycommons.net/artifacts/12281693/generally_faster_-_the_economic_impact_of_generative_ai/

Miller, S. (2018). *Dual use science and technology, ethics and weapons of mass destruction*. Springer. https://doi.org/10.1007/978-3-319-92606-3

Misselhorn, C. (2018). *Grundfragen der Maschinenethik*. Reclam.

Mitcham, C. (2014). Agency in humans and in artifacts: A contested discourse. In P. Kroes & P.-P. Verbeek (Eds.), *The moral status of technical artefacts* (Philosophy of Engineering and Technology, Vol. 17) (pp. 11–29). Springer. https://doi.org/10.1007/978-94-007-7914-3

OECD. (2024). *OECD digital economy outlook 2024 (Volume 1): Embracing the technology frontier*. OECD. https://doi.org/10.1787/a1689dc5-en

Rath, J., Ischi, M., & Perkins, D. (2014). Evolution of different dual-use concepts in international and national law and its implications on research ethics and governance. *Science and Engineering Ethics, 20*(3), 769–790. https://doi.org/10.1007/s11948-014-9519-y

Scholz, R. W., Bartelsman, E. J., Diefenbach, S., Franke, L., Grunwald, A., Helbing, D., Hill, R., Hilty, L., Hoejer, M., Klauser, S., Montag, C., Parycek, P., Prote, J. P., Renn, O., Reichel, A., Schuh, G., Steiner, G., & Pereira, G. V. (2018). Unintended side effects of the digital transition: European scientists' messages from a proposition-based expert round table. *Sustainability, 10*(6), 2001. https://doi.org/10.3390/su10062001

Sevini, F., Colpo, P., Lequarré, A.-S., Charatsis, Ch., Barrero, J., Michel, Q., Janssens, W. A. M., & Novau, X. A. (2018, January). *Emerging dual-use technologies and global supply chain compliance*. Presentation at the IAEA Symposium on International Safeguards 2018, Vienna, Austria. Retrieved March 31, 2025, from https://www.researchgate.net/publication/330258872_EMERGING_DUAL-USE_TECHNOLOGIES_AND_GLOBAL_SUPPLY_CHAIN_COMPLIANCE

Standing, G. (2018). Taskers in the precariat: Confronting an emerging dystopia. In E. Paus (Ed.), *Confronting dystopia: The new technological revolution and the future of work* (pp. 115–133). Cornell University Press. https://doi.org/10.7591/9781501719868-007

Susskind, R. E., & Susskind, D. (2015). *The future of the professions: How technology will transform the work of human experts* (1st ed.). Oxford University Press. https://doi.org/10.1093/oso/9780198713395.001.0001

Turkle, S. (2007). Authenticity in the age of digital companions. *Interaction Studies, 8*(3), 501–517. https://doi.org/10.1075/is.8.3.11tur

UN & International Labour Organization. (2024). *Mind the AI divide: Shaping a global perspective on the future of work*. Retrieved March 31, 2025, from https://www.ilo.org/publications/major-publications/mind-ai-divide-shaping-global-perspective-future-work

UNESCO. (2024). *Ethics of artificial intelligence: The recommendation*. Retrieved March 31, 2025, from https://www.unesco.org/en/artificial-intelligence/recommendation-ethics

Ungerer, A. (2019, October 16). Als Vater war er gleichermassen grossartig und furchtbar. *Die Zeit Magazin*. Retrieved March 31, 2025, from https://www.zeit.de/zeit-magazin/2019/43/aria-ungerer-tomi-ungerer-tochter-literatur

Verbeek, P.-P. (2005). *What things do: Philosophical reflections on technology, agency, and design*. Penn State University Press. https://doi.org/10.1007/s11948-007-9010-0

Verbeek, P.-P. (2011). *Moralizing technology: Understanding and designing the morality of things*. University of Chicago Press. https://doi.org/10.7208/9780226852904

Verbeek, P.-P. (2015). Designing the public sphere: Information technologies and the politics of mediation. In L. Floridi (Ed.), *The onlife manifesto: Being human in a hyperconnected era* (pp. 208–219). Springer. https://doi.org/10.1007/978-3-319-04093-6_21

Virt, G. (2007). Damit Menschsein Zukunft hat. In G. Marschuetz & G. M. Prueller-Jagenteufel (Eds.), *Theologische Ethik im Einsatz für eine humane Gesellschaft*. Echter.

Weizenbaum, J. (1976). *Computer power and human reason: From judgment to calculation*. W. H. Freeman.

Willcocks, L. (2020). Robo-apocalypse cancelled? Reframing the automation and future of work debate. *Journal of Information Technology, 35*(3), 286–302. https://doi.org/10.1177/0268396220925830

Winner, L. (1980). Do artifacts have politics? *Daedalus, 109*(1), 121–136. http://www.jstor.org/stable/20024652

World Health Organization. (n.d.). *Dual use research of concern (DURC)*. Retrieved March 31, 2025, from https://www.who.int/csr/durc/en/

CHAPTER 2

Creation or Destruction of Paid Jobs?

Abstract This chapter addresses the question of whether digital transformation and DS will lead to a reduction of paid professional tasks for humans. Historically, this fear always came up amid a technological boost. So far, more technology has always led to more human jobs. Comparing digital transformation and DS with earlier technology-based eras of change, it turns out, among others:

- Unlike earlier boosts of innovation, digital transformation and DS are, first of all, not about making work easier for humans, but replacing humans.
- Digital transformation and DS aspire that less to no human input will be needed.
- Digital transformation and DS concern all professional jobs (e.g., robot surgeons, robot lawyers, etc.), not just paid professional tasks requiring low or no qualification.

Keywords Digital transformation • Data-based systems (DS) • Paid professional tasks • Technology • Human jobs • Reduction • Human history • Decrease

P. G. Kirchschlaeger, *Ethics and the Digital Transformation of Human Work*, https://doi.org/10.1007/978-3-031-98704-5_2

Without doubt, digital transformation and the use of DS create new jobs (Al Naqbi et al., 2024; Balsmeier & Woerter, 2019; Bessen, 2019; Capgemini & Prognos, 2018; German Federal Ministry of Labour and Social Affairs, 2017; Gordon, 2016; Green, 2023; Weber, 2021), open new horizons for innovative business models and economic systems as well as for new forms of work (Jensen & Koch, 2015; Zimmermann, 2015). Digital transformation and DS could contribute to technology-based approaches to solving social and ecological problems. This dynamic will cause more productivity—especially in generative DS (Al Naqbi et al., 2024)—and economic growth (Capgemini and Prognos, 2018; Green, 2023). It will require an adaptation of the labor market (Arntz et al., 2017; Weber, 2021). Technology-based support in business processes leads to disruptions (Schumpeter, 1972, pp. 134–142) and, at the same time, relief for humans and creates new space for other tasks (Dengler & Matthes, 2015; Eichhorst et al., 2016; Elstner et al., 2016; Oerder, 2016; Staab & Nachtwey, 2016).

Will DS substitute, limit, or lighten human labor (Hasenbein, 2023; Von Richthofen et al., 2021)?

> The digitalization of our economies can be a blessing. Robots can make work lighter, solve labor shortages and make our life easier. But there are also grounds for concern. Potentially, there are implications for demand of labor, quality and nature of work, workers' representation and quality and role of social dialogue. Some are concerned that digitalization will be a steamroller dramatically changing our world of work, completely unchecked. (Metzger, 2019, p. 76)

The main concern is whether many more paid professional tasks for humans will disappear than will be created (Betschon, 2015; Brynjolfsson & McAfee, 2011; Cowen, 2013; Ford, 2009, 2015; Frey & Osborne, 2017; McKinsey Global Institute, 2017; Ohly, 2019; Schweighofer, 2016), which would significantly distinguish this transformation from other boosts of innovation.

The responses to this inquiry (Mondolo, 2021) in the academic discourse can be clustered into the following five lines of argumentation:

1. compensation
2. mutual reinforcement
3. spoiling
4. domination
5. substitution

2.1 Compensation

One line of argumentation recognizes the disruption and substitution of some professional tasks by DS while emphasizing that DS will require humans to fulfill the remaining professional tasks (Tokunova et al., 2024) or new professional tasks (Dauth et al., 2021) complementing the performance of DS (African Union Commission [AUC] & Organisation for Economic Co-operation and Development [OECD], 2021; Blanas et al., 2019; Carbonell, 2022; Casilli, 2019; De Bellefonds et al., 2024; Dosi et al., 2021; Elgohary & Abdel-Aziz, 2022; Ernst et al., 2019; Fleming, 2018; Green, 2024; Gregory et al., 2016; Huang et al., 2019; Dengler & Matthes, 2015; McKinsey & Company, 2024; Sharfaei, 2024) and that technological progress will discover new business opportunities (Schmalzried et al., 2023), achieve productivity gains (Brynjolfsson et al., 2023) via automation (Chui et al., 2023; PricewaterhouseCoopers [Pwc], 2017, p. 4), elevate the demand of consumers amid the availability of personalized and higher-quality products and services because of DS (Pwc, 2017, p. 4), and cause economic growth (Acemoglu, 2024; Carlsson-Szlezak & Swartz, 2024; Chui et al., 2023; Cui et al., 2024; McFee, 2024) including a positive impact on human labor (Azpiazu Arrieta & Bayón Pérez, 2022; Brynjolfsson et al., 2018; Dauth et al., 2017; Dixon et al., 2021; European Parliament, 2018; Georgieff & Hyee, 2021; Han et al., 2024; Klenert et al., 2022; Lan & Cui, 2024; Ling et al., 2023; Liu et al., 2024; Montobbio et al., 2021; Qiu & Yin, 2023; World Economic Forum [WEF] 2025; Ye & Cai, 2024), on human wages (Felten et al., 2019), and other global economic challenges—for example, "health care, education, infrastructure, housing" (Pollin, 2018).

> Automation does indeed substitute for labor – as it is typically intended to do. However, automation also complements labor, raises output in ways that lead to higher demand for labor, and interacts with adjustments in labor supply. [...] Changes in technology do alter the types of jobs available and what those jobs pay. (Autor, 2015, p. 5)

It needs to be taken into consideration, though, that—even if new jobs are created—they are not created in the places where paid jobs are destroyed by the digital transformation and DS (Acemoglu & Restrepo, 2020).

This line of argumentation "compensation" enjoys the support of other reasons, namely, the current status quo of use of DS in the private sector. Looking at the concrete realities (Haertwig et al., 2023) in concrete companies—so the argument goes—a lack of substantial use of DS does not indicate a reduction of human labor by DS (Arntz et al., 2020; Giering, 2021).

At the same time, one must take into account: "It's abundantly clear to all that artificial intelligence, big data analytic, and advanced robotics make it possible for machines to take on tasks that once required a person to do them" (Knickrehm, 2019, p. 99).

Besides that, no one would question that substantial changes occur in the economy or the labor market.

> In the wake of these dynamic changes, inappropriate and ubiquitous digitalization may lead to social exclusion due to increased unemployment resulting from the mismatch between the supply of competencies and the demand for them, which is likely to disturb the economic balance and have a negative effect on the economic security of the population. (Kosala et al., 2022, p. 7)

Another reasoning supporting the line of argumentation "compensation" relies on a historical insight: Never in human history have innovation-based shifts led to a decrease in jobs (Hessler, 2016).

> One of the greatest ironies of this period in history is that, just as technology remakes our world, the need to maintain the human dimension of our work, and a company's sense of its social responsibility, is growing at an equally rapid pace. Harmonizing economic growth with the protection of human rights is one of the greatest challenges we face today. (Office of the United Nations High Commissioner for Human Rights 2000)

Concerning this pattern of argumentation, doubts arise whether every innovation-based transformation happens in the same way. First, one should consider the complexity of technological progress.

> Technological change is a complex, nonlinear, and costly process. It is not homogenous, and it is here defined broadly to take into account the forms of innovation that affect the quantity and nature of jobs in different ways. It is reflected in the discoveries of new scientific principles and inventions; in the implementation of new or significantly improved products and new

> industries (product innovations); and in new production techniques, organizational routines, and business models (process innovations); as well as in the wide diffusion of these innovations in the economy. (Nuebler, 2018, p. 47)

Second, one should keep in mind the potential uniqueness of this technology-based shift compared to earlier innovation-based eras. It is an intellectual temptation to compare and to equate this technological epoch of change to earlier times of technology-based revolution (Brynjolfsson & McAfee, 2014, pp. 2–12). Most of the discourse on digital transformation and DS does the same (Cevasco et al., 2019; Kaiser & Malanowski, 2019; Lenz, 2018). On a metaethical level, yielding to this intellectual temptation provokes three fundamental hermeneutic doubts: Why should it be the same as before? There are no reasons why one could automatically assume that. This is, for example, the underlying assumption of an affirmation like this one:

> Computers and other digital advances are doing for mental power – the ability to use our brains to understand and shape our environments – what the steam engine and its descendants did for muscle power. They're allowing us to blow past previous limitations and taking us into new territory. [...] So, a vast and unprecedented boost to mental power should be a great boost to humanity, just as earlier boost to physical power so clearly was. (Brynjolfsson & McAfee, 2014, pp. 7–8)

In contrast, theoretically it could also be possible that this technology-based transformation limits humanity or degenerates it. The idea here is to show that there are no reasons why the former should prevail over the latter. Just because something was the case with the Industrial Revolution, this will not automatically repeat itself with the digital transformation and DS. In fact, following such a uniform line of argumentation implies an understanding of human history as a linear progress always and eternally following the same logics. With history in general and especially in the case of history of science and technology, it can be shown that this is not the case (Kuhn, 1962; Laudan, 1977).

Why should every innovation-based transformation lead to the same pattern of consequences, including benefiting from them if they are positive and overcoming them if they are negative?

> Digitalization is going to bring with it some thorny challenges. This in itself should not be too surprising or alarming; even the most beneficial developments have unpleasant consequences that must be managed. The Industrial Revolution was accompanied by soot-filled London skies and horrific exploitation of child labor. What will be their modern equivalents? (Brynjolfsson & McAfee, 2014, p. 10)

The argument continues by pointing out that these challenges were met on the occasion of the Industrial Revolution and therefore that will be also the case with the negative consequences of the digital transformation and DS. This seems to be an unfounded line of argumentation because, on the one hand, child labor and environmental pollution are still unacceptable parts of the value chain of digital transformation and DS today. On the other, sharing the position that humans were able to get rid of the ethically negative sides of the Industrial Revolution, this does not automatically mean that humans will also be able to liberate themselves from the ethical downsides of digital transformation and DS. "We stand on the brink of a technological revolution that will fundamentally alter the way we live, work, and relate to one another. In its scale, scope, and complexity, the transformation will be unlike anything humankind has experienced before" (Schwab, 2016).

Why should something innovative, of all things, lead to the same method and path and reach similar patterns of result? "It [the second machine age] is an inflection point in the right direction" (Brynjolfsson & McAfee, 2014, p. 11). Could it not be possible that all the ethically negative effects listed by Erik Brynjolfsson and Andrew McAfee themselves make it that this innovation-based transformation is actually taking us in the *wrong direction*—namely, by destroying humanity and planet Earth—and that humans need to change that direction immediately? Just because something is new neither means that it is leading humanity in the right direction nor that it is ethically legitimate.

Beyond that, progress in itself cannot be an end from an ethical perspective.

> All progress is progress toward an end, but there can be no progress in the formation, expression, and evaluation of ends themselves, only either their circumscription or proliferation. […] to treat progress as an end-in-itself is usually to disavow actual ends – whether for equity, for diversity, for increasing gross domestic product, for more reported satisfactions on the happiness index, whatever the end may be – and by so disavowing them so too to

> relinquish the terms without which any notion of progress at all is finally unintelligible. (Carrico, 2013, p. 50)

Finally, one needs to acknowledge: "The future is a process, not a destination" (Sterling, 2004).

The line of argumentation "compensation" results in the following recommendation: "This comes to bear on policymaking, as the authorities should design interventions aimed at improving the quality of human capital and at producing skills that are complemented, rather than substituted, by technological change" (Mondolo, 2021, p. 1059). This process will be defined by the following rationale:

> Although broader economic effects can be complex, labor demand is more likely to fall for tasks that are close substitutes for capabilities of ML [Machine Learning], whereas it is more likely to increase for tasks that are complements for these systems. Each time an ML system crosses the threshold where it becomes more cost-effective than humans on a task, profit-maximizing entrepreneurs and managers will increasingly seek to substitute machines for people. This can have effects throughout the economy, boosting productivity, lowering prices, shifting labor demand, and restructuring industries. (Brynjolfsson & Mitchell, 2017, p. 1531)

One of the consequences of this development for humans would be to change the business-model accordingly, which will lead to an increase in paid professional tasks for humans (Brooks et al., 2020). Amid a boost of efficiency based on the use of DS causing a reduction of paid professional tasks for humans, why should a company change this already more efficient and therefore economically profitable business model risking to lose again the just-achieved gains of efficiency while knowing that paid professional tasks for humans normally are more expensive than the fulfillment of professional tasks by DS—even if certain compromises in quality have to be conceded?

What we can learn methodologically from this line of argumentation "compensation" for the further ethical analysis of digital transformation of human work and the impact of the use of DS on human labor is that one should not discuss the increase or decrease of entire paid professions but rather focus on specific skills (Felten et al., 2018) and on specific professional tasks in order to take into account the unique technological character of DS.

> Jobs typically consist of a number of distinct but interrelated tasks. In most cases, only some of these tasks are likely to be suitable for ML, and they are not necessarily the ones that were easy to automate with previous technologies. For instance, [...] we find that a ML system can be trained to help lawyers classify potentially relevant documents for a case but would have a much harder time interviewing potential witnesses or developing a winning legal strategy. (Brynjolfsson & Mitchell, 2017, p. 1533)

Furthermore, it needs to be considered that economic growth does not need to cause an increase in employment or human wages (McFee, 2024). Though, from an ethical standpoint, the follow-up question always has to be raised about who benefits from positive economic developments.

2.2 Mutual Reinforcement

Another position embraces that DS will enable humans to create new work-opportunities (Acemoglu & Restrepo, 2019; Agrawal et al., 2023; Autor et al., 2022; Fischer, 2023; Gondlach & Regneri, 2023), boost the productive capacities (Chui et al., 2023; Henkin & Zangrilli, 2024; Jaffe et al., 2024; Willcocks, 2020; Yu & Qi, 2024), and augment the cognitive functions of humans (Brynjolfsson et al., 2023; WEF, n.d.; PwC, 2017, p. 4; Cevasco et al., 2019). At the same time, DS will benefit afterwards from the interaction with humans (Walkowiak, 2023) who then will already be "boosted humans," leading to a continuous mutual reinforcement of humans' and DS' productivity.

> Companies see the biggest performance gains when humans and smart machines collaborate. People are needed to train machines, explain their outputs, and ensure their responsible use. AI, in turn, can enhance humans' cognitive skills and creativity, free workers from low-level tasks, and extend their physical capabilities. (Wilson & Daugherty, 2018, p. 4)

2.3 Spoiling

One should not avoid what the aspired boost of the productivity of humans looks like—among others, the negative impact of the "travail du clic" (Casilli, 2019) on the work-situation for humans (Carbonell, 2022; Casilli, 2019) as well as on human wages (Bessen et al., 2023; Carbonell, 2022; Casilli, 2019; Korinek, 2023; Korinek & Suh, 2024; Meier, 2024; Restrepo, 2023).

> The platforms maximize profits through ownership and control of the technological apparatus, protected by patents and other intellectual property rights, and by the exploitation of labor through tasking and unpaid work. As platform-based tasking expands, it will be appreciated just how isolated the precariat is in this zone, in constant competition with one another. The atomization drives down wages and transfers costs, risks, and uncertainty onto the precariat. So far, at least, taskers have had minimal means or opportunities to coalesce. These forms of labor intensify pressure to commodify all aspects of life. Intensifying self-exploitation is a sad way for the precariat to respond to adversity. It is how those experiencing declining wages and living standards cover up the decline, for a while. But when sweating spreads to taskers, the threat to wages and working conditions for all workers grows. The precariat's vulnerability today is a threat to all of us tomorrow. (Standing, 2018, p. 133)

2.4 Domination

Humanity has fought for centuries for human rights and for their implementation and realization, human rights in their universality are ethically justifiable (Kirchschlaeger, 2013; Kirchschlaeger, 2016), and therefore should not be given up amid, for example, technology-based possibilities of totalitarian surveillance, control (United Nations [UN] and International Labor Organization 2024, p. 10), and exploitation (Irani & Silberman, 2013) of humans at the workplace (Carbonell, 2022; Sainato, 2020).

> One paradigmatic example of the potential negative impact of the massive deployment of AI and robotics technologies on workers is the treatment of people employed in the Amazon warehouses in the US, as described in some journalistic reports. Working in a highly automated environment that dictates a tight and inflexible performance schedule, embedded in a thick network of AI-powered cameras, sensors, wearables, and constantly under the observation of the technology-enhanced invisible eye of their employer, workers report feeling imprisoned and surveyed continuously, subject to the risk of physical and psychological harm, provided with poor medical and psychological support, deprived of any meaningful social interaction, and under the constant threat of losing their job. (Santoni de Sio, 2024a, p. 6)

The fundamental basis for these violations of human rights represents the untransparent "algorithmic control": "They use information control as a tool to maintain an uncontrolled power over workers" (Santoni de Sio, 2024a, p. 6). This is combined with the loss of "meaningful human control" over DS (Santoni de Sio & Van Den Hoven, 2018).

> Workers and other people interacting with AI systems may end up having only a superficial control over its functioning while losing a substantive capacity and opportunity to have the system respond to their relevant intentions and plans. This may be due to an excessive complexity or opacity in the system functioning, an insufficient capacity and motivation to engage in meaningful interaction or a bad design of the human-machine interaction. The lack of sufficient meaningful human control may create so-called "responsibility gaps", situations in which no human being can legitimately be held responsible for an unwanted outcome of the human-machine interaction. (Santoni de Sio, 2024a, p. 413)

At the same time, in the case of success, the credits go to DS, not to the participating and contributing, if not leading, humans—provoking a "achievement gap" (Danaher & Nyholm, 2020).

Beyond that, transforming the formula "if you are not paying for the product, you are the product" (Christiaens, 2022) (you pay for the service with your data, not with money) to "if you're not paying for the product, you are the worker" (Santoni de Sio, 2024b, pp. 74–94) allows to highlight the robbery of multinational technology-companies consisting of the sale of online activities of humans to third parties without spending anything in salaries for this work (Santoni de Sio, 2024a, p. 416).

> Companies owning big digital online platforms are exploitative in two ways: they exploit the work of public agencies like research institutes, schools, etc., by extracting an unreciprocated gain from the cognitive skills and knowledge created by them; and they exploit the free labor of users of the platforms who produce the traffic and content that make the platforms successful and economically valuable. (Santoni de Sio, 2024a)

The humane design of human work (Bankins & Formosa, 2023; Cirillo et al., 2021; Costantino et al., 2021; Fleming, 2018; Furendal & Jebari, 2023; Goos, 2018; Hacker, 2022; Kauffeld & Maier, 2020; Krings & Frey, 2024; Latos et al., 2024; Mirbabaie et al., 2021; Nitsch et al., 2024; Petrat, 2021; Stofile et al., 2023; Tausch & Kluge, 2023; Vogt, 2021) and the respect of human rights at the workplace and on the labor market are necessary from an ethical perspective. This does not change under high pressure due to DS.

> Digital technologies create new opportunities to organize work in a more flexible manner, allowing workers to become self-employed and develop

> new activities and income opportunities. This enhances people's choices; however, new institutions and regulations are needed to provide social security and ensure workers' rights. (Nuebler, 2018)

These—due to digital transformation of human work and the impact of DS on human labor—already existing harsh realities for humans still fulfilling paid professional tasks (Acemoglu & Johnson, 2024) with their potential to even grow in the future (Hirsch-Kreinsen et al., 2018; Hirsch-Kreinsen & Karacic, 2019; Osadchaya et al., 2024; Ponce del Castillo, 2023) do decrease the argumentative power of the line of argumentation "compensation" due to its normative problems.

> The problem is that the rich traditions of moral thought that guide human relationships have no equivalent when it comes to robot-human-interaction (…). We face a future in which robots will test the boundaries of our ethical and legal frameworks with increasing audacity. (Nourbakhsh, 2023)

2.5 Substitution

The line of argumentation "substitution" consists in emphasizing that the digital transformation of human work and the impact of the use of DS will cause a massive reduction of paid professional tasks for humans. This line of argumentation states that DS will provoke or cause a disruption, and DS will take away the paid professional tasks from humans and will replace them (Acemoglu et al., 2022). An illustrative example for the line of argumentation "substitution" is automated checkout systems in grocery stores in the present that substitute human cashiers as a means for the end of an increase in efficiency due to the lower costs for the DS than for cashiers on the payroll.

The main counterargument against the line of argumentation "substitution" is a historical observation: When there was a technology-based transformational era in human history, there was always the fear that technological progress would push humans out of labor. Never in human history did this fear become true. Innovation-based shifts led to an increase in jobs, mostly because of the economic growth initiated by technological innovations and technology-based transformation of the economy. Therefore, it is necessary to examine this counterargument to see if it is still valid also for digital transformation of human work and the impact of the use of DS on human labor. This critical analysis follows in the next chapter.

Before this point is examined more profoundly, a moment should be dedicated to a self-critical meta reflection: So far, the deliberations in this chapter have focused on the *rather empirical question* if there are reasons plausibilizing that the digital transformation of human work and the impact of the use of DS on human labor encompasses or will consist in a compensation, a mutual reinforcement, a spoiling, a domination, or a substitution. The spotlight did not turn yet to the *normative question* of whether one ought to strive for compensation, mutual reinforcement, spoiling, domination, or substitution via the digital transformation of human work and the use of DS. Admittedly, while discussing the lines of argumentation "spoiling" and "domination," more than a hint was provided that both paths would be ethically problematic. But regarding these two lines of argumentation as well as regarding all lines of argumentation, a more profound ethical assessment will follow in Chap. 5, if necessary. This ethical in-depth study would be necessary—and this represents also the intention behind the current focus on the rather empirical question of whether there are reasons plausibilizing that the digital transformation of human work and the impact of the use of DS on human labor encompasses or will consist in a compensation, a mutual reinforcement, a spoiling, a domination, or a substitution—if rational reasons can be presented which show that one or some of these lines of argumentation are realistic. If the opposite should be the case, namely, rational reasons could be presented which show that one or some of these lines of argumentation are not realistic, then that ethical assessment of that line respectively those lines of argumentation would lose its necessity.

References

Acemoglu, D. (2024). *The simple macroeconomics of AI* (NBER Working Paper No. 32487). National Bureau of Research. https://doi.org/10.3386/w32487

Acemoglu, D., & Johnson, S. (2024). Learning from Ricardo and Thompson: Machinery and labor in the early Industrial Revolution and in the age of artificial Intelligence. *Annual Review of Economics, 16*(1), 597–621. https://doi.org/10.1146/annurev-economics-091823-025129

Acemoglu, D., & Restrepo, P. (2019). Automation and New Tasks: How technology displaces and reinstates labor. *The Journal of Economic Perspectives, 33*(2), 3–30. https://doi.org/10.1257/jep.33.2.3

Acemoglu, D., & Restrepo, P. (2020). Robots and Jobs: Evidence from US Labor Markets. *Journal of Political Economy, 128*(6), 2188–2244. https://doi.org/10.1086/705716

Acemoglu, D., Autor, D., Hazell, J., & Restrepo, P. (2022). Artificial intelligence and jobs: Evidence from online vacancies. *Journal of Labor Economics, 40*(S1), 293–340. https://doi.org/10.1086/718327

Agrawal, A., Gans, J., & Goldfarb, A. (2023). *The Turing transformation: Artificial intelligence, intelligence augmentation, and skill premiums* (NBER Working Paper No. 31767). National Bureau of Economic Research. https://doi.org/10.3386/w31767

Al Naqbi, H. A., Bahroun, Z., & Ahmed, V. (2024). Enhancing work productivity through generative artificial intelligence: A comprehensive literature review. *Sustainability, 16*(3), 1166. https://doi.org/10.3390/su16031166

Arntz, M., Gregory, T., & Zierahn, U. (2017). Revisiting the risk of automation. *Economics Letters, 159*, 157–160. https://doi.org/10.1016/j.econlet.2017.07.001

Arntz, M., Gregory, T., & Zierahn, U. (2020). Digitalisierung und die Zukunft der Arbeit. *Wirtschaftsdienst, 100*(S1), 41–47. https://doi.org/10.1007/s10273-020-2614-6

AUC/OECD. (2021). *Africa's development dynamics 2021: Digital transformation for quality jobs.* OECD Publishing. https://doi.org/10.1787/0a5c9314-en

Autor, D. H., Chin, C., Salomons, A., & Seegmiller, B. (2022). New frontiers: The origins and content of new work, 1940–2018 (NBER Working Paper No. 30389). National Bureau of Economic Research. https://doi.org/10.3386/w30389

Autor, D. H. (2015). Why are there still so many jobs? The history and future of workplace automation. *The Journal of Economic Perspectives, 29*(3), 3–30. https://doi.org/10.1257/jep.29.3.3

Azpiazu Arrieta, G., & Bayón Pérez, B. P. (2022). Tendencias laborales y el futuro del trabajo por medio de la robotización, digitalización e inteligencia artificial en España. *Revista Razón Crítica, 12*, 1–15. https://doi.org/10.21789/25007807.1805

Balsmeier, B., & Woerter, M. (2019). Is this time different? How digitalization influences job creation and destruction. *Research Policy, 48*(8), 103765. https://doi.org/10.1016/j.respol.2019.03.010

Bankins, S., & Formosa, P. (2023). The ethical implications of Artificial Intelligence (AI) for meaningful work. *Journal of Business Ethics, 185*(4), 725–740. https://doi.org/10.1007/s10551-023-05339-7

Bessen, J. (2019). Automation and jobs: When technology boosts employment. *Economic Policy, 34*(100), 589–626. https://doi.org/10.1093/epolic/eiaa001

Bessen, J., Goos, M., Salomons, A., & van den Berge, W. (2023, February 6). Automatic reaction: What happens to workers at firms that automate? *The Review of Economics and Statistics.* Advance online publication. Retrieved March 31, 2025, from https://scholarship.law.bu.edu/faculty_scholarship/584

Betschon, S. (2015, August 21). Automatisierung: Ein neues Maschinenzeitalter. *Neue Zuercher Zeitung*. Retrieved March 31, 2025, from https://www.nzz.ch/gesellschaft/lebensart/gesellschaft/ein-neues-maschinenzeitalter-1.18599135

Blanas, S., Gancia, G., & Lee, S. Y. (2019). Who is afraid of machines? *Economic Policy, 34*(100), 627–690. https://doi.org/10.1093/epolic/eiaa005

Brooks, C., Gherhes, C., & Vorley, T. (2020). Artificial intelligence in the legal sector: Pressures and challenges of transformation. *Cambridge Journal of Regions, Economy and Society, 13*(1), 135–152. https://doi.org/10.1093/cjres/rsz026

Brynjolfsson, E., Li, D., & Raymond, L. (2023). *Generative AI at work* (NBER Working Paper No. 31161). National Bureau of Economic Research. https://doi.org/10.3386/w31161

Brynjolfsson, E., & McAfee, A. (2011). *Race against the machine: How the digital revolution is accelerating innovation, driving productivity, and irreversibly transforming employment and the economy*. Digital Frontier Press.

Brynjolfsson, E., & McAfee, A. (2014). *The second machine age: Work, progress, and prosperity in a time of brilliant technologies*. W. W. Norton & Company.

Brynjolfsson, E., & Mitchell, T. (2017). What can machine learning do? Workforce implications. *Science, 358*(6370), 1530–1534. https://doi.org/10.1126/science.aap8062

Brynjolfsson, E., Mitchell, T., & Rock, D. (2018). What can machines learn, and what does it mean for occupations and the economy? *AEA Papers and Proceedings, 108*, 43–47. https://doi.org/10.1257/pandp.20181019

Capgemini & Prognos. (2018). *Gesellschaft 5.0. Implikationen der Digitalisierung für ausgewählte Lebensfelder*. Retrieved March 31, 2025, from https://www.prognos.com/de/projekt/gesellschaft-50

Carbonell, J. S. (2022). *Le futur du travail*. Éditions Amsterdam. https://doi.org/10.4000/travailemploi.12495

Carlsson-Szlezak, P., & Swartz, P. (2024, August 15). *Why AI will not lead to a world without work*. WEF. Retrieved March 31, 2025, from https://www.weforum.org/stories/2024/08/why-ai-will-not-lead-to-a-world-without-work/

Carrico, D. (2013). Futurological discourses and posthuman terrain. *Existenz: An International Journal in Philosophy, Religion, Politics, and the Arts, 8*(2), 47–63.

Casilli, A. A. (2019). *En attendant les robots : enquête sur le travail du clic*. Éditions du Seuil.

Cevasco, L. J., Corvalán, J. G., & Le Fevre Cervini, E. M. (2019). *Intelligenza artificiale e lavoro. Costruire un nuovo paradigma occupazionale*. Edizioni di Comunità.

Christiaens, T. (2022). *Digital working lives: Worker Autonomy and the gig Economy*. Rowman & Littlefield Publishers.

Chui, M., Hazan, E., Roberts, R., Singla, A., Smaje, K., Sukharevsky, A., Yee, L., & Zemmel, R. (2023, June 14). *The economic potential of generative AI: The next productivity frontier*. McKinsey & Company. Retrieved March 31, 2025 from https://www.mckinsey.com/capabilities/mckinsey-digital/ourinsights/the-economic-potential-of-generative-ai-the-next-productivity-frontier#introduction

Cirillo, V., Rinaldini, M., Staccioli, J., & Virgillito, M. E. (2021). Technology vs. workers: The case of Italy's Industry 4.0 factories. *Structural Change and Economic Dynamics, 56*, 166–183. https://doi.org/10.1016/j.strueco.2020.09.007

Costantino, F., Falegnami, A., Fedele, L., Bernabei, M., Stabile, S., & Bentivenga, R. (2021). New and emerging hazards for health and safety within digitalized manufacturing systems. *Sustainability, 13*(19), 10948. https://doi.org/10.3390/su131910948

Cowen, T. (2013). *Average is over: Powering America beyond the age of the great stagnation*. Dutton.

Cui, K. Z., Demirer, M., Jaffe, S., Musolff, L., Peng, S., & Salz, T. (2024). The productivity effects of generative AI: Evidence from a field experiment with GitHub Copilot. *An MIT Exploration of Generative AI*. https://doi.org/10.21428/e4baedd9.3ad85f1c

Danaher, J., & Nyholm, S. (2020). Automation, work and the achievement gap. *AI and Ethics, 1*(3), 227–237. https://doi.org/10.1007/s43681-020-00028-x

Dauth, W., Findeisen, S., Suedekum, J., & Woessner, N. (2017, September 1). *German robots: The impact of industrial robots on workers*. CEPR Discussion Paper No. DP12306. Centre for Economic Policy Research. Retrieved March 31, 2025, from https://papers.ssrn.com/sol3/papers.cfm?abstract_id=3039031

Dauth, W., Findeisen, S., Suedekum, J., & Woessner, N. (2021). The adjustment of labor markets to robots. *Journal of the European Economic Association, 19*(6), 3104–3153. https://doi.org/10.1093/jeea/jvab012

De Bellefonds, N., Charanya, T., Franke, M. R., Apotheker, J., Forth, P., Grebe, M., Luther, A., De Laubier, R., Lukic, V., Martin, M., Nopp, C., & Sassine, J. (2024, October 24). Where's the value in AI? *BCG Global*. Retrieved March 31, 2025, from https://www.bcg.com/publications/2024/wheres-value-in-ai

Dengler, K., & Matthes, M. (2015). Folgen der Digitalisierung für die Arbeitswelt: Substituierbarkeitspotenziale von Berufen in Deutschland [IAB-Forschungsbericht 11/2015]. Institut für Arbeitsmarkt- und Berufsforschung. Retrieved March 31, 2025, from https://doku.iab.de/forschungsbericht/2015/fb1115.pdf

Dixon, J., Hong, B., & Wu, L. (2021). The robot revolution: Managerial and employment consequences for firms. *Management Science, 67*(9), 5586–5605. https://doi.org/10.1287/mnsc.2020.3812

Dosi, G., Piva, M., Virgillito, M., & Vivarelli, M. (2021). Embodied and disembodied technological change: The sectoral patterns of job-creation and job-destruction. *Research Policy, 50*(4), 104-199. https://doi.org/10.1016/j.respol.2021.104199

Eichhorst, W., Hinte, H., Rinne, U., & Tobsch, V. (2016). *Digitalisierung und Arbeitsmarkt: Aktuelle Entwicklungen und sozialpolitische Herausforderungen* (IZA Standpunkte, No. 85). Institut zur Zukunft der Arbeit (IZA). Retrieved March 31, 2025, from https://docs.iza.org/sp85.pdf

Elgohary, E. M., & Abdel-Aziz, R. O. (2022). The effect of digital transformation on employment in Egypt: An applied study using ARDL model. *The Electronic Journal of Information Systems in Developing Countries, 89*(1). https://doi.org/10.1002/isd2.12240

Elstner, S., Feld, L. P., & Schmidt, C. M. (2016). Bedingt abwehrbereit – Deutschland im digitalen Wandel. *Wirtschaftspolitische Blätter, 63*(2), 287–308.

Ernst, E., Merola, R., & Samaan, D. (2019). Economics of Artificial Intelligence: Implications for the future of work. *IZA Journal of Labor Policy, 9*(1). https://doi.org/10.2478/izajolp-2019-0004

European Parliament. (2018). *The impact of new technologies on the labour market and the social economy*. Directorate-General for Parliamentary Research Services & Dachs, European Parliament. https://doi.org/10.2861/68448

Felten, E. W., Raj, M., & Seamans, R. (2018). A method to link advances in artificial intelligence to occupational abilities. *AEA Papers and Proceedings, 108*, 54–57. https://doi.org/10.1257/pandp.20181021

Felten, E.W., Raj, M., & Seamans, R. (2019, September 8). The occupational impact of artificial intelligence: Labor, skills, and polarization. *NYU Stern School of Business*. Retrieved March 31, 2025, from https://ssrn.com/abstract=3368605

Fischer, F. (2023). Future collaboration between humans and AI. In I. Knappertsbusch & K. Gondlach (Eds.), *Work and AI 2030* (pp. 21–28). Springer. https://doi.org/10.1007/978-3-658-40232-7_3

Fleming, P. (2018). Robots and Organization Studies: Why robots might not want to steal your job. *Organization Studies, 40*(1), 23–38. https://doi.org/10.1177/0170840618765568

Ford, M. (2009). *The lights in the tunnel: Automation, accelerating technology and the economy of the future*. Acculant Publishing.

Ford, M. (2015). *Rise of the robots: Technology and the threat of a jobless future*. Basic Books.

Frey, C. B., & Osborne, M. A. (2017). The future of employment: How susceptible are jobs to computerisation? *Technological Forecasting and Social Change, 114*, 254–280. https://doi.org/10.1016/j.techfore.2016.08.019

Furendal, M., & Jebari, K. (2023). The future of work: Augmentation or stunting? *Philosophy & Technology, 36*(2). https://doi.org/10.1007/s13347-023-00631-w

Georgieff, A., & Hyee, R. (2021). *Artificial intelligence and employment: New cross-country evidence* (OECD Social, Employment and Migration Working Paper No. 265). OECD Publishing. https://doi.org/10.1787/c2c1d276-en

German Federal Ministry of Labour and Social Affairs. (2017). *Weissbuch: Arbeiten 4.0. Arbeit weiterdenken*. Retrieved March 31, 2025, from https://www.bmas.de/DE/Service/Publikationen/Broschueren/a883-weissbuch.html

Giering, O. (2021). Künstliche Intelligenz und Arbeit: Betrachtungen zwischen Prognose und betrieblicher Realität. *Zeitschrift Für Arbeitswissenschaft, 76*(1), 50–64. https://doi.org/10.1007/s41449-021-00289-0

Gondlach, K. A., & Regneri, M. (2023). The ghost of German angst: Are we too skeptical for AI development? In I. Knappertsbusch & K. Gondlach (Eds.), *Work and AI 2030* (pp. 3–10). Springer. https://doi.org/10.1007/978-3-658-40232-7_1

Goos, M. (2018). The impact of technological progress on labour markets: Policy challenges. *Oxford Review of Economic Policy, 34*(3), 362–375. https://doi.org/10.1093/oxrep/gry002

Gordon, R. J. (2016). *The rise and fall of American growth: The U.S. standard of living since the Civil War*. Princeton University Press.

Green, A. (2023). Artificial intelligence and jobs: No signs of slowing labour demand (yet). In *OECD Employment Outlook 2023: Artificial intelligence and the labour market* (pp. 102–127). OECD Publishing. https://doi.org/10.1787/9c86de40-en

Green, A. (2024). *Artificial intelligence and the changing demand for skills in the labour market* (OECD Artificial Intelligence Papers, No. 14). OECD Publishing. https://doi.org/10.1787/88684e36-en

Gregory, T., Salomons, A., & Zierahn, U. (2016). Racing with or against the machine? Evidence from Europe. *SSRN Electronic Journal*. https://doi.org/10.2139/ssrn.2815469

Hacker, W. (2022). Arbeitsgestaltung bei Digitalisierung. *Zeitschrift Für Arbeitswissenschaft, 76*(1), 90–98. https://doi.org/10.1007/s41449-022-00302-0

Haertwig, C., Sapronova, A., & Sigmund, N. (2023). Wer profitiert von der Digitalisierung? Deutschlandweite Follow-up-Studie zu Entwicklungsverläufen und Unterschiedseffekten bei Beschäftigten verschiedener Industriebranchen und Tätigkeitsfelder. *Zeitschrift Für Arbeitswissenschaft, 77*(4), 578–592. https://doi.org/10.1007/s41449-023-00386-2

Han, Y., Yang, J., Ying, L., & Niu, Y. (2024). The impact of corporate digital transformation on labor employment. *Finance Research Letters, 60*, 104888. https://doi.org/10.1016/j.frl.2023.104888

Hasenbein, M. (2023). Wandel der Arbeits- und Organisationswelt durch Künstliche Intelligenz. In M. Hasenbein (Ed.), *Mensch und KI in Organisationen* (pp. 35–55). Springer. https://doi.org/10.1007/978-3-662-66375-2_4

Henkin, D., & Zangrilli, C. (2024). Enhancement, not elimination: AI's impact on jobs in the indirect tax domain. *Communications of the Association for Information Systems, 55*, 846–853. https://doi.org/10.17705/1cais.05532

Hessler, M. (2016). Zur Persistenz der Argumente im Automatisierungsdiskurs. *Aus Politik und Zeitgeschichte, 66*(18–19), 17–24. Retrieved March 31, 2025, from https://www.bpb.de/apuz/225690/zur-persistenz-der-argumente-im-automatisierungsdiskurs

Hirsch-Kreinsen, H., Ittermann, P., & Niehaus, J. (2018). *Digitalisierung industrieller Arbeit: die Vision Industrie 4.0 und ihre sozialen Herausforderungen* (2nd ed.). Nomos.

Hirsch-Kreinsen, H., & Karacic, A. (2019). *Autonome Systeme und Arbeit: Perspektiven, Herausforderungen und Grenzen der Künstlichen Intelligenz in der Arbeitswelt*. transcript. https://doi.org/10.1515/9783839443958

Huang, M., Rust, R., & Maksimovic, V. (2019). The feeling economy: Managing in the next generation of artificial intelligence (AI). *California Management Review, 61*(4), 43–65. https://doi.org/10.1177/0008125619863436

Institut für Arbeitsmarkt- und Berufsforschung. (2015). *Folgen der Digitalisierung für die Arbeitswelt: Substituierbarkeitspotenziale von Berufen in Deutschland* [IAB-Forschungsbericht 11/2015]. Retrieved March 31, 2025, from https://doku.iab.de/forschungsbericht/2015/fb1115.pdf

Irani, L. C., & Silberman, M. S. (2013). Turkopticon. In W. E. Mackay, S. Brewster, & S. Bødker (Eds.), *Proceedings of the SIGCHI conference on human factors in computing systems* (pp. 611–620). ACM. https://doi.org/10.1145/2470654.2470742

Jaffe, S., Shah, N. P., Butler, J., Farach, A., Cambon, A., Hecht, B., Schwarz, M., & Teevan, J. (2024). *Generative AI in real-world workplaces – Microsoft Research*. Microsoft Research. Retrieved March 31, 2025, from https://www.microsoft.com/en-us/research/publication/generative-ai-in-real-world-workplaces/

Jensen, B., & Koch, M. (2015). *Mensch und Maschine: Roboter auf dem Vormarsch? Folgen der Automatisierung für den Schweizer Arbeitsmarkt*. Deloitte.

Kaiser, O. S., & Malanowski, N. (2019). *Smart Data und Künstliche Intelligenz: Technologie, Arbeit, Akzeptanz* (Working Paper Forschungsförderung, No. 136). Hans-Böckler-Stiftung. Retrieved March 31, 2025, from http://hdl.handle.net/10419/216056

Kauffeld, S., & Maier, G. W. (2020). Digitalisierte Arbeitswelt. *Gruppe Interaktion Organisation Zeitschrift Für Angewandte Organisationspsychologie (GIO), 51*(1), 1–4. https://doi.org/10.1007/s11612-020-00508-y

Kirchschlaeger, P. G. (2013). *Wie können Menschenrechte begründet werden? Ein für religiöse und säkulare Menschenrechtskonzeptionen anschlussfähiger Ansatz* (ReligionsRecht im Dialog (Vol. 15)). LIT-Verlag.

Kirchschlaeger, P. G. (2016). How can we justify human rights? *International Journal of Human Rights and Constitutional Studies, 4*(4), 313–329. https://doi.org/10.30965/23642807-00601006

Klenert, D., Fernández-Macías, E., & Antón, J. (2022). Do robots really destroy jobs? Evidence from Europe. *Economic and Industrial Democracy, 44*(1), 280–316. https://doi.org/10.1177/0143831x211068891

Knickrehm, M. (2019). How will AI change work? Here are five schools of thought. In T. H. Davenport, E. Brynjolfsson, A. McAfee, & H. J. Wilson (Eds.), *Artificial intelligence: The insights you need from Harvard Business Review* (pp. 99–108). Harvard Business Press.

Korinek, A. (2023). Scenario planning for an A(G)I future. *Finance and Development Magazine.* Retrieved March 31, 2025, from https://www.imf.org/en/Publications/fandd/issues/2023/12/Scenario-Planning-for-an-AGI-future-Anton-korinek

Korinek, A., & Suh, D. (2024). *Scenarios for the transition to AGI* (Working Paper 32255). National Bureau of Economic Research. https://doi.org/10.3386/w32255

Kosala, M., Jelonek, M., & Gorzelany-Dziadkowiec, M. (2022). Developing competencies for the future. In S. Mazur (Ed.), *Industrial revolution 4.0: Economic foundations and practical implications* (pp. 7–29). Routledge.

Krings, B., & Frey, P. (2024). Human-centered approaches in working environments needs ethical reflection – Experiences from KARL. *Zeitschrift Für Arbeitswissenschaft, 78*(3), 349–358. https://doi.org/10.1007/s41449-024-00436-3

Kuhn, T. S. (1962). *The structure of scientific revolutions.* University of Chicago Press.

Lan, S., & Cui, M. (2024). Enterprise digital transformation and employment: Spillover effect within supply chains. *Finance Research Letters, 67*, 105928. https://doi.org/10.1016/j.frl.2024.105928

Latos, B., Buckhorst, A., Kalantar, P., Bentler, D., Gabriel, S., Dumitrescu, R., Minge, M., Steinmann, B., & Guhr, N. (2024). Time autonomy in personnel planning: Requirements and solution approaches in the context of intelligent scheduling from a holistic organizational perspective. *Zeitschrift für Arbeit swissenschaft, 78*, 277–298. https://doi.org/10.1007/s41449-024-00432-7

Laudan, L. (1977). *Progress and its problems: Towards a theory of scientific growth.* University of California Press.

Lenz, F. (2018). *Digitalisierung und Beschäftigung: Ein Ende ohne Arbeit oder Arbeit ohne Ende?* (Argumente zu Marktwirtschaft und Politik, No. 141). Stiftung Marktwirtschaft. Retrieved March 31, 2025, from http://hdl.handle.net/10419/177820

Ling, X., Luo, Z., Feng, Y., Liu, X., & Gao, Y. (2023). How does digital transformation relieve the employment pressure in China? Empirical evidence from the national smart city pilot policy. *Humanities and Social Sciences Communications, 10*(1). https://doi.org/10.1057/s41599-023-02131-w

Liu, J., Chen, K., & Lyu, W. (2024). Embracing artificial intelligence in the labour market: The case of statistics. *Humanities and Social Sciences Communications, 11*(1). https://doi.org/10.1057/s41599-024-03557-6

McFee, I. (2024). Feature article: How GenAI will change the world economy. *Economic Outlook, 48*(3), 39–46. https://doi.org/10.1111/1468-0319.12767

McKinsey & Company. (2024, November 15). *How to get more value from your new tech? Five technologists weigh in.* Retrieved March 31, 2025, from https://www.mckinsey.com/about-us/new-at-mckinsey-blog/how-to-getmore-value-from-your-new-tech-five-experts-weigh-in

McKinsey Global Institute. (2017). *Jobs lost, jobs gained: What the future of work will mean for jobs, skills, and wages.* McKinsey & Company. Retrieved March 31, 2025, from https://www.mckinsey.com/~/media/McKinsey/Industries/Public%20and%20Social%20Sector/Our%20Insights/What%20the%20future%20of%20work%20will%20mean%20for%20jobs%20skills%20and%20wages/MGI-Jobs-Lost-Jobs-Gained-Executive-summary-December-6-2017.pdf

Meier, D. (2024). *The future of work: Ethical evaluation of the change of human labor in the context of advancing automation.* Nomos.

Metzger, E. (2019). Opportunities and challenges of digitalization for work and employment, working conditions and social dialogue. In D. Buerkardt, H. Kohler, N. Kreuzkamp, & J. Schmid (Eds.), *Smart Factory und Digitalisierung: Perspektiven aus vier europäischen Ländern und Regionen = Perspectives from four European countries and regions* (1st ed., pp. 63–77). Nomos.

Mirbabaie, M., Bruenker, F., Frick, N. R. J. M., & Stieglitz, S. (2021). The rise of artificial intelligence – Understanding the AI identity threat at the workplace. *Electronic Markets, 32*(1), 73–99. https://doi.org/10.1007/s12525-021-00496-x

Mondolo, J. (2021). The composite link between technological change and employment: A survey of the literature. *Journal of Economic Surveys, 36*(4), 1027–1068. https://doi.org/10.1111/joes.12469

Montobbio, F., Staccioli, J., Virgillito, M. E., & Vivarelli, M. (2021). Robots and the origin of their labour-saving impact. *Technological Forecasting and Social Change, 174*, 121122. https://doi.org/10.1016/j.techfore.2021.121122

Nitsch, V., Rick, V., Kluge, A., & Wilkens, U. (2024). Human-centered approaches to AI-assisted work: The future of work? *Zeitschrift Für Arbeitswissenschaft, 78*(3), 261–267. https://doi.org/10.1007/s41449-024-00437-2

Nourbakhsh, I. R. (2023, June 22). The coming robot dystopia: all too inhuman. *Foreign Affairs.* Retrieved March 31, 2025, from https://www.foreignaffairs.com/coming-robot-dystopia

Nuebler, I. (2018). New technologies, innovation, and the future of jobs. In E. Paus (Ed.), *Confronting Dystopia: The new technological revolution and the future of work* (pp. 46–75). Cornell University Press. https://doi.org/10.7591/9781501719868

Oerder, K. (2016). Mitbestimmung 4.0. Der Wandel der Arbeitswelt als Chance für mehr Beteiligung. *WISO Direkt, 24*, 1–4.

Office of the UN High Commissioner for Human Rights (2000). *Business and human rights: A progress report*. Retrieved March 31, 2025, from https://www.ohchr.org/Documents/Publications/BusinessHRen.pdf

Ohly, L. (2019). *Ethik der Robotik und der Künstlichen Intelligenz*. (Theologisch-Philosophische Beiträge zu Gegenwartsfragen 22). Peter Lang.

Osadchaya, E., Marder, B., Yule, J. A., Yau, A., Lavertu, L., Stylos, N., Oliver, S., Angell, R., De Regt, A., Gao, L., Qi, K., Zhang, W. Z., Zhang, Y., Li, J., & AlRabiah, S. (2024). To ChatGPT, or not to ChatGPT: Navigating the paradoxes of generative AI in the advertising industry. *Business Horizons, 67*(5), 571–581. https://doi.org/10.1016/j.bushor.2024.05.002

Petrat, D. (2021). Artificial intelligence in human factors and ergonomics: An overview of the current state of research. *Discover Artificial Intelligence, 1*(1), 3. https://doi.org/10.1007/s44163-021-00001-5

Pollin, R. (2018). Expanding job opportunities through global green growth. In E. Paus (Ed.), *Confronting Dystopia: The new technological revolution and the future of work* (pp. 76–93). Cornell University Press. https://doi.org/10.7591/9781501719868

Ponce del Castillo, A. (2023). Generative AI, generating precariousness for workers? *AI & Society, 39*(5), 2601–2602. https://doi.org/10.1007/s00146-023-01719-9

PwC. (2017). *Sizing the prize: What's the real value of AI for your business and how can you capitalise?* Retrieved March 31, 2025, from https://www.pwc.com/gx/en/issues/data-and-analytics/publications/artificial-intelligence-study.html

Qiu, H., & Yin, Z. (2023). Digital transformation, information search, and women's high-quality employment. *China Finance and Economic Review, 12*(3), 45–64. https://doi.org/10.1515/cfer-2023-0015

Restrepo, P. (2023). *Automation: Theory, evidence, and outlook* (Working Paper 31910). National Bureau of Economic Research. https://doi.org/10.3386/w31910

Sainato, M. (2020, February 5). I'm not a robot: Amazon workers condemn unsafe, grueling conditions at warehouse. *The Guardian*. Retrieved March 31, 2025, from https://www.theguardian.com/technology/2020/feb/05/amazon-workers-protest-unsafe-grueling-conditions-warehouse

Santoni de Sio, F. (2024a). Artificial intelligence and the future of work: Mapping the ethical issues. *The Journal of Ethics, 28*(3), 407–427. https://doi.org/10.1007/s10892-024-09493-6

Santoni de Sio, F. (2024b). *Human freedom in the age of AI*. Routledge.
Santoni de Sio, F., & Van Den Hoven, J. (2018). Meaningful human control over autonomous systems: A philosophical account. *Frontiers in Robotics and AI, 5*. https://doi.org/10.3389/frobt.2018.00015
Schmalzried, D., Hurst, M., Wentzien, M., & Graeser, M. (2023). Analyse der Rolle Künstlicher Intelligenz für eine menschenzentrierte Industrie 5.0. *HMD Praxis Der Wirtschaftsinformatik, 60*(6), 1143–1155. https://doi.org/10.1365/s40702-023-01001-y
Schumpeter, J. A. (1972). *Kapitalismus, Sozialismus und Demokratie*. Francke.
Schwab, K. (2016, January 14). *The Fourth Industrial Revolution: What it means and how to respond*. World Economic Forum. Retrieved March 31, 2025, from https://www.weforum.org/stories/2016/01/the-fourth-industrial-revolution-what-it-means-and-how-to-respond/
Schweighofer, J. (2016). Zur Befreiung des Menschen von mühevoller Arbeit und Plage durch Maschinen, Roboter und Computer – Auswirkungen der Digitalisierung auf die Arbeitsmärkte. *Wirtschaft und Gesellschaft, 42*(2), 219–255. https://EconPapers.repec.org/RePEc:clr:wugarc:y:2016v:42i:2p:219
Sharfaei, S. (2024). Robots et emploi: revue de la littérature. *Revue Internationale Du Travail, 163*(2), 301–325. https://doi.org/10.1111/ilrf.12310
Staab, P., & Nachtwey, O. (2016). Die Digitalisierung der Dienstleistungsarbeit. *Politik und Zeitgeschichte, 18–19*, 24–31.
Standing, G. (2018). Taskers in the precariat: Confronting an emerging dystopia. In E. Paus (Ed.), *Confronting Dystopia: The new technological revolution and the future of work* (pp. 115–133). Cornell University Press.
Sterling, B. (2004, June 11). *The singularity: Your future as a black hole*. The Long Now Foundation. Retrieved March 31, 2025, from http://longnow.org/seminars/02004/jun/11/the-singularity-your-future-as-a-black-hole/#
Stofile, P., De Braine, R., & Dhanpat, N. (2023). Experiencing work in the Fourth Industrial Revolution: A qualitative study on work identity and new ways of work. *EUREKA Social and Humanities, 4*, 3–12. https://doi.org/10.21303/2504-5571.2023.002955
Tausch, A., & Kluge, A. (2023). Teaming with robots – Principles of sustainable humane work design. *Zeitschrift Für Arbeitswissenschaft, 77*(4), 478–489. https://doi.org/10.1007/s41449-023-00389-z
Tokunova, A., Zvonar, V., Polozhentsev, D., Pavlova, V., & Fedoruk, O. (2024). Economic consequences of artificial intelligence and labor automation: Employment recovery, transformation of labor markets, and dynamics of social structure in the context of digital transformation. *Financial Engineering, 2*, 1–12. https://doi.org/10.37394/232032.2024.2.1

UN & International Labour Organization. (2024). *Mind the AI divide: shaping a global perspective on the future of work.* Retrieved March 31, 2025, from https://www.ilo.org/publications/major-publications/mind-ai-divide-shaping-global-perspective-future-work

Vogt, J. (2021). Where is the human got to go? Artificial intelligence, machine learning, big data, digitalisation, and human–robot interaction in Industry 4.0 and 5.0. *AI & Society, 36*(3), 1083–1087. https://doi.org/10.1007/s00146-020-01123-7

Von Richthofen, G., Gümüsay, A. A., & Send, H. (2021). Künstliche Intelligenz und die Zukunft von Arbeit. In R. Altenburger & R. Schmidpeter (Eds.), *CSR und Künstliche Intelligenz. Management-Reihe Corporate Social Responsibility.* Springer Gabler. https://doi.org/10.1007/978-3-662-63223-9_19

Walkowiak, E. (2023). Task-interdependencies between generative AI and workers. *Economics Letters, 231*, 111315. https://doi.org/10.1016/j.econlet.2023.111315

Weber, E. (2021). Digitalisierung und Arbeitsmarkt: Wirkungen und Herausforderungen. In J. Muckenhuber, H. J. Griesbacher, & L. Zilian (Eds.), *Disruption der Arbeit? Zu den Folgen der Digitalisierung im Dienstleistungssektor* (1st ed., pp. 21–30). Campus.

WEF. (2025, January). *Future of jobs report 2025. Insight Report.* World Economic Forum. Retrieved March 31, 2025, from https://www.weforum.org/publications/the-future-of-jobs-report-2025/

WEF. (n.d.). *Where will AI take us? 6 experts discuss.* World Economic Forum. Retrieved March 31, 2025 from https://www.weforum.org/videos/ai-6-experts/

Willcocks, L. (2020). Robo-apocalypse cancelled? Reframing the automation and future of work debate. *Journal of Information Technology, 35*(4), 286–302. https://doi.org/10.1177/0268396220925830

Wilson, J. H., & Daugherty, P. R. (2018). Collaborative intelligence: Humans and AI are joining forces. *Harvard Business Review, 96*(4), 114–123.

Ye, R., & Cai, X. (2024). Digital transformation, gender discrimination, and female employment. *Systems, 12*(5), 162. https://doi.org/10.3390/systems12050162

Yu, J., & Qi, C. (2024). The impact of generative AI on employment and labor productivity. *Review of Business, 44*(1), 53–67.

Zimmermann, K. F. (2015). *Nur noch Roboter und Selbstausbeutung? Über die Herausforderungen und Chancen der neuen Welt der Arbeit* (IZA Standpunkte 1 Nr. 80). Institut zur Zukunft der Arbeit (IZA). Retrieved March 31, 2025, from https://docs.iza.org/sp80.pdf

CHAPTER 3

The Uniqueness of Digital Transformation and Data-Based Systems (DS)

Abstract In this chapter, unique characteristics of digital transformation and DS are identified. They indicate that this time will be different: 30 differentiators from earlier technology-based eras of change hollow the main argument against the potential of massive reduction of paid professional tasks for humans, namely, that never in human history innovation-based shifts have led to a decrease in jobs. Combined with other considerations, this analysis results in the assessment that likely many more paid professional tasks for humans will disappear than will be created due to digital transformation and DS, which significantly distinguishes this transformational era from other boosts of innovation.

Keywords Digital transformation • Data-based systems (DS) • Unique characteristics • Differentiators • Technology-based era of change • Massive reduction of paid professional tasks • Boosts of innovation • Transformational era

"Much of the history of technology and work can be seen as a race between technologies that create new types of work and automation technologies that replace existing ones" (Frey & Osborne, 2024, p. 11). This characterization is certainly valid for digital transformation and data-based systems

P. G. Kirchschlaeger, *Ethics and the Digital Transformation of Human Work*, https://doi.org/10.1007/978-3-031-98704-5_3

(DS) as well. The question is whether more jobs are created than destroyed or are more jobs replaced than generated due to digital transformation and the use of DS. A response to this question can be sought in the assessment of similarities and differences between this technology-based shift compared to earlier ones: If this technology-based transformation is like previous innovative eras of transition, then probably also its impact on human work follows the pattern of past technological transformational periods which can be summarized with the formula: more technologies—more jobs!

If digital transformation and DS are unlike what humanity has seen so far as technology-based change, their effect on human labor could differ as well.

For these reasons, now the question to investigate is whether there are differentiators which make digital transformation and DS unique compared to previous eras of technological change in the history of humanity.

First, unlike earlier technology-based eras of change (Hessler, 2016), digital transformation and DS are, first of all, not about making work easier, more interesting, or different for humans, but replacing humans in the value chain for reasons of cost-efficiency. For example, when companies install automatic checkout systems in supermarkets, they do not aim at making the professional job of a cashier easier, as was the case for the farmer when a plough was replaced by a tractor. They would rather replace cashiers and remove them from the payroll in order to gain from cost-efficiency.

Second, even if DS do not reach the same level of quality as humans or are not able to fulfill all tasks humans would be able because of the unattainability of certain domains of human intelligence for DS, due to the hard competition of a globalized economy—as hard as never before because as globalized as never before—DS will be preferred over humans. (Andrew McAfee, co-founder of the Initiative on the Digital Economy at the Massachusetts Institute of Technology, MIT, states: "I don't think a lot of employers are going to be willing to pay a lot of people for what they are currently doing" (McAfee, as cited by Gross, 2014)).

Third, DS, and especially self-learning DS, are made on purpose aspiring to ensure that less to no human input will be needed (Clarke & Whittlestone, 2022; Floridi, 2015, p. 53; Hirsch-Kreinsen & Karacic, 2019; Neuburger & Fiedler, 2020; Russell, 2021).

Fourth, digital transformation and DS concern all professional jobs (e.g., robot surgeons, robot lawyers, robot judges, etc.) (Alimardani,

2024; Amankwah-Amoah et al., 2024; Andrei et al., 2024; Atchley et al., 2024; Batista et al., 2024; Baumgartner et al., 2024; Bender, 2024; Bergstrom & Ruediger, 2024; Bond et al., 2024; Boussioux et al., 2024; Broecke, 2023; Chan & Tsi, 2024; Clusmann et al., 2023; Coetzee & Veldsman, 2022; Cook et al., 2024; Corvalán, 2018; Crowe et al., 2024; Cui et al., 2024; Davenport et al., 2019; De Luca Picione et al., 2024; Demirci et al., 2024; Deng et al., 2024; Duffy, 2018; Erickson, 2024; Fear & Gleber, 2023; Ferioli, 2019; Fogleman et al., 2024; Ghimire et al., 2024; Grewal et al., 2024; Hartmann, 2023; Hartmann et al., 2017, p. 5; Hermida, 2024; Horchi & Ait Soudane, 2024; Huber et al., 2023; Humbsch et al., 2024; Islam & Islam, 2024; Jede et al., 2024; Juca-Maldonado, 2023; Kan et al., 2024; Koegel et al., 2023; Krusche et al., 2023; Kuziemski & Misuraca, 2020; Langford et al., 2024; Lee, 2024; Li et al., 2022; Lopezosa et al., 2024; McKinsey & Company, 2024; Montejo et al., 2024; Mueller et al., 2024; Nordlinger et al., 2024; Nowrozy, 2024; Noy & Zhang, 2023; Osadchaya et al., 2024; Owen-Smith, 2024; Pelizzari et al., 2024; Pieterson et al., 2023; Poisat et al., 2024; Prasad & De, 2024; Pressman et al., 2024; Ransbotham et al., 2024; Reddy, 2024; Romera-Paredes et al., 2023; Ruediger et al., 2024; Ruh et al., 2024; Ruehr et al., 2023; Sako, 2024; Schmidt, 2023; Séïde, 2024; Stark & Pais, 2020; Stowasser, 2023; Susskind & Susskind, 2015; Varnosfaderani & Forouzanfar, 2024; Wachter & Brynjolfsson, 2023; Watermeyer et al., 2023; Wenz & Ebener, 2024; Xia et al., 2024) – not just paid professional tasks requiring low or no qualification that were affected by earlier technology-based transformations (Levy & Murnane, 2004). "Increasingly capable machines [...] will take on more and more non-routine tasks; and so the intuition that there will always tasks left that only humans can perform will prove to be ill-founded" (Susskind & Susskind, 2015, p. 394).

Fifth, digital transformation and DS are strengthened by globalization and vice versa (Kaashoek et al., 2024; Knight & Khan, 2022; Wilhelms, 2018, p. 3), which also has a unique speed-up effect on the changes caused by digital transformation and DS in the professional world. On the one hand, digital transformation and DS are fueled by global access to datasets (big data playing a constitutive role for digital transformation and DS) (Atuahene et al., 2022; Zuboff, 2019) instead of only local, national, or regional data. On the other hand, globalization is fueled by the high-quality, simplified, and cost-free technological instruments and communication as well as information possibilities thanks to digital transformation and DS.

Sixth, digital transformation embraces, in a unique way, several technology-based innovations with impacts on human labor (Martinelli et al., 2021), for example, digitalization, automation, mechanization, and robots (Carbonero et al., 2018; Compagnucci et al., 2019; De Vries et al., 2020; Dixon et al., 2021; Fu et al., 2020; Klenert et al., 2022; DS Raj & Seamans, 2019).

Seventh, digital transformation and DS happen at a much higher speed in shorter intervals than during earlier technology-based eras of change, so society as a whole does not have enough time to adapt before DS have taken away the paid professional tasks of humans.

Eighth, technology-based innovations spread globally as fast as never before. "This new paradigm turns out to be even more rapid in its diffusion and more flexible in its adoption" (Barbieri et al., 2019, p. 2). The speed of its diffusion (Taddy, 2018) is even further increased by a specific technological characteristic of DS: "Unlike solutions to tasks mastered by humans, many solutions to tasks automated by ML can be disseminated almost instantly worldwide" (Brynjolfsson & Mitchell, 2017, p. 1534).

Ninth, DS transform the professional and the private lives of humans, not just their professional lives, which intensifies the penetration of technology into human existence and therefore accelerates its progress and, as a result, its impact on human labor.

Tenth, the nature of digital transformation and DS—above all, being nourished directly by data of the same humans—enhances the acceptance-rate and the willingness to get accustomed to new technological applications fostering the impact of this technology-based change on human labor.

Eleventh, digital transformation and the use of DS entail the end of striving for full employment because of the "great decoupling" (Brynjolfsson & McAffee, 2011) of the increase in productivity based on technological progress and the slowdown of total employment since 2011. "Productivity is at record levels, innovation has never been faster, and yet at the same time, we have a falling median income, and we have fewer jobs" (Rotman, 2013). In other terms, "a unit of wealth can now typically be created with far fewer workers than in previous industrial eras, owing to low marginal costs (tending towards zero) associated with nonrival and nonexcludable digital goods" (Chalmers et al., 2020, p. 1030; Goldfarb & Tucker, 2019; Schwab, 2016). This new phenomenon can be illustrated by the company WhatsApp having "only 55 employees but over 450 million users when it sold to Facebook in 2014 for USD 19b" (Chalmers et al., 2020, p. 1030).

Twelfth, for the first time in human history, the revenue and the profit of a company can be enormous and, at the same time, the number of employees extremely small (Metzler, 2017).

Thirteenth, platform-economy increases efficiency but not the incentives and the motivation to create paid professional tasks for humans. On the contrary, they lower the number of paid professional tasks.

> We live in an era where fewer than 10% of the world's public companies account for more than 80% of all profits. […] Ten years ago, banks and energy companies dominated the top ten. Today, it's technology companies, with US computer company Apple in the number one spot. (Gray, 2017)

Seven of the world's ten biggest companies make money with platforms (Gray, 2017). Besides offering multidimensional and diverse business opportunities for their owners, platforms raise concerns from an ethical perspective. "The transparency and global reach of digital platforms is triggering a downward price spiral. Internet platforms have potential for modern slavery. […] The cheapest provider of a service is awarded the contract" (Hofstetter, 2017, p. 80). Normative frameworks as part of the economic system are under attack or torn down on these global marketplaces. Human rights are neglected with just a click, the former being something for which humanity has been fighting for centuries in order to protect the survival and a life as human—a life with human dignity. Digital transformation and the use of DS also consist in a devaluation of human labor forcing humans into "multi-jobbing"—enabled by platforms like Uber and Airbnb, for example. "The people who hire themselves out as multi-jobbers and market themselves via online platforms are the cause of unimaginable sales and profits – but not for themselves, but for the platform operators" (Hofstetter, 2017, p. 83, translated by the author)

Fourteenth, investments and funding are attracted by this stimulus of technology-based innovations resulting in high profits with a low number of paid professional tasks for humans as never before in human history, which leads even to an intensification and reinforcement of this development towards an economy consisting of economic growth with a decrease in paid professional tasks for humans.

Fifteenth, the response to the question of whether humans will still be allowed to fulfill a paid professional task does not depend on the skills of the humans but rather on the definition and the description of the job (Ford, 2015) which are dominated though, as mentioned above, by digital transformation and DS as well as by cost-efficiency in hyper-competitive globalized markets.

Sixteenth, based on the core consequence of digital transformation—*ever fewer humans will directly participate and partake in a more efficient and effective value chain* (Kirchschlaeger, 2016; Kirchschlaeger, 2017), and fewer humans involved in research, science, and technology decrease the diversity of the involved humans, which provokes a lower probability of innovation creating new paid professional tasks for humans.

Seventeenth, if the vast majority of humans does not have the possibility to earn enough for their survival and for a life with human dignity by fulfilling paid professional tasks and, therefore, the trust of humans that they are born with something that makes the difference, "the ability to work, to learn, and to earn money" (Smith, 2013) is fading away, it is not only ethically unacceptable but this development also provokes a decrease in innovation because humans will have to focus on the essentials—their survival. This way, another source of new paid professional tasks for humans dries out.

Eighteenth, due to this fast technology-based shift, elements of the unknown and uncertain represent the new reality as well as the only certainty, which makes it impossible for humans to prepare themselves adequately by upskilling and by educating themselves further so paid professional tasks for humans float away.

Nineteenth, the nature, the characteristics, and the profiles of paid professional tasks are not static. They change or are changed. Due to digital transformation and the comprehensive penetration of the economy and of the labor market by DS, this technology-based change also extends to the design, the structure, and the organization of paid professional tasks: paid professional tasks are adapted in a way that they accommodate the specifications and the abilities of DS (Acemoglu & Restrepo, 2020; Wilson & Daugherty, 2018) in order to become as cost-efficient as possible, not to the capacities of humans.

Twentieth, while DS further develop and improve in fulfilling professional tasks thanks to the intense interaction with humans as well as because of the enormous amount of data generated by the constant presence of DS in private and professional lives of humans, humans—by relying on DS in more and more areas of private and professional lives—degenerate in certain domains of their intelligence so their professional competitiveness against DS on the labor market declines. If one looks, for example, at science and academia—a human occupation which requires cognitive capacities—one sees that first researchers and academics start to cite LLMs (e.g., ChatGPT)—the DS which are constantly, systematically, and constitutively violating copyrights and intellectual property rights by being trained with ideas, thoughts, creations, and words from humans without neither remunerating nor citing them as well as by refraining from referencing the sources—in their scientific articles and academic papers, this is not only perverse but also a sign of human degeneration.

> It doesn't make sense when you consider what a citation is supposed to be. We cite for two reasons. One is that the source carries some epistemic authority. This isn't necessarily because of the author – it could be because of our trust (or at least the performance of trust) in the practice of peer review. Or maybe we're citing a statistic produced by an official body. The other reason for citing is that we want to credit the source. Neither of these applies to AI. Chatbots have no epistemic authority – or at least shouldn't if we're invested in maintaining some distinction between scholarship and slop. And chatbots aren't moral agents nor have ownership of the text they output, so they don't need to be credited. (Narayanan, 2025)

Twenty-first, the fulfillment of these professional tasks adapted for DS not only fits better with machines but can also be rigorously surveilled with DS, provoking in humans more stress and less happiness (Becker, 2017). Suffering from both is the competitiveness of humans against DS in the fight for paid professional tasks.

Twenty-second, the striving to make humans perfect, to which so-called "human enhancement" contributes in its various forms, is in fact destabilizing humans based on the alleged notion of representing a being of deficiency which needs to be improved. Framing humans in this negative way puts humans under pressure, in particular at their workplace (Sandel, 2007). This increases the advantages of DS over humans on the labor market.

Twenty-third, the striving for perfection in the course of digital transformation and the use of DS—with an understanding of perfection based on DS and its essential characteristics, not on humans—favors DS over humans. "However, perfected technology also leads to something like an inverse anthropology: the more perfect, the more determined things ideally are or should be, the more noticeable the human factors, the degrees of freedom of human behavior, appear as a source of irritation in technical scenarios. The individual or collectively 'the' user is then the irregular element: unpredictable, imperfect. He (or she) is a poor fit" (Alpsancar et al., 2023, p. 9, translated by the author).

Twenty-fourth, humans can be manipulated with or by DS (OECD, 2023) which was not possible in the history of humankind with technology-based innovations during earlier transformational eras of technological progress. All the time, data is stolen from humans and sold to the highest bidder. This constant violation of human rights to privacy and data protection—forming a massive attack on the freedom of all humans—causes the possession of a vast amount of data about humans, enabling one to know humans better than humans know themselves with or by DS. This opens the door to economic and political manipulation as well as disinformation. With DS, one can know—to use a metaphor—exactly which piano keys to press to make the music play, in other words, to get people to buy or choose or vote the way the hands controlling DS want them to (Kirchschlaeger, 2024). Several democratic decisions have already been manipulated so far with the help of DS. Take the 2016 presidential election in the USA: It has been proven that "Facebook" (now: Meta) sold off user data records. The same thing happened with Brexit (Andrzejewski, 2023). This means that dictators and totalitarian regimes can influence elections and votes in democracies. Manipulation and disinformation lead to the destabilization of democratic countries. It is now possible with "social media" to destabilize within a few hours a peaceful, functioning, and wealthy country (e.g., the UK far-right riots 2024 (Cadwalladr, 2024)). The rapidly developing technical possibilities for disinformation and manipulation of people through large language models such as ChatGPT open new horizons in this regard. It can be assumed that, for example, ChatGPT will further intensify the phenomenon of fake news, which is already so devastating for democracies, with "deep fakes." It is now possible with "Generative AI" to destroy a politician with a "Hollywood-quality" deep fake (Verma & Zakrzewski, 2024).

At the same time, quality journalism as a pillar of democracy will come under even greater economic and political pressure. This is because media channels can be filled with texts from, e.g., ChatGPT at a low cost while to this generative DS it does not matter at all if the generated information is false or not because the criterion of truthfulness does not play any role. The only purpose generative DS serve is to fill the communication- or information-channel with some content. If that content is "fake news" is not even irrelevant to generative DS like ChatGPT. For example, if one looks for a response to a specific question in ethics, it can be the case that generative DS goes so far to invent the academic article in ethics as well as the ethics-journal in which this fake article is published.

Moreover, economic manipulation affects humans as consumers. Based on the enormous amount of data from humans and using DS, one can precisely trigger someone else to buy anything one wants.

This manipulative potential in the political as well as in the economic sphere opens the possibility of pushing forward faster and intensifying without critical ethical examination the replacement of humans by DS.

Twenty-fifth, digital transformation and DS evolve in a globalized scientific, technological, and economic environment so a technology-based enthusiasm and hype as well as a blind belief in technological progress pushing DS into the labor market and pushing humans out of the labor market resonates in a global echo chamber.

Twenty-sixth, societal, political, or legal opposition against certain technology-based changes and their impact on human labor is lost in space because—while protests against effects of previous technology based transitions in human history happened, for example, in front of a concrete automated factory—in the case of digital transformation and DS, it is hard to know where to start as this technology-based wave seems to happen everywhere and nowhere at the same time and therefore it seems to be hard to master it.

Twenty-seventh, generative DS or large language models build a currently already existing technology which illustrates in an exemplary way the amplification of domains (Capraro et al., 2024; Mohammed & Skibniewski, 2023) where DS can simulate the fulfillment of functions by humans (Barenkamp, 2023; Eloundou et al., 2023; Jošt et al., 2024; Kanbach et al., 2024; Leyh & Schaeffer, 2024; Orchard & Tasiemski, 2023; Schoenberger & Beinke, 2023; Seufert & Meier, 2023) while amplifying the violations of human rights by or through DS as well (e.g., discrimination (Larsson et al., 2024) and the violations of intellectual property rights (Chesterman, 2024)). This amplification of domains reenforces the massive reduction of professional tasks for humans by DS.

> Generative AI is technological innovation with endless potential to transform businesses and society. Despite its immense potential, there are still limitations that need to be acknowledged and present across all industries, including but not limited to privacy, ethical dimensions, and issues related to data ownership. It is only by overcoming these limitations will businesses be able to effectively implement generative AI into their operations. (Ooi et al., 2023, p. 26)

It can be assumed that other areas of cognitive skills of humans will follow. The effect with respect to paid professional tasks and paid jobs already caused by self-learning machines will be further amplified by "super-DS" representing DS which surpass human intelligence by far in certain domains of human intelligence.

Just as a side note: At the same time, one needs to remain critical and not conceptually exaggerate reality. The starting point for this thought is "functional equivalences" (Ohly, 2019, p. 18)—indicating that something is producing the same result in the same way.

> I see the greatest danger in robotics in the fact that we can replace human abilities with the abilities of robots and falsely conclude that robots have our abilities. In comparison: You can move faster with a bicycle than on foot. But it doesn't mean that bicycles have feet that go faster. When robots take over human tasks it doesn't even mean that they themselves engage in those activities. When they replace human decisions, it doesn't mean that they decide themselves. And when they replace human feelings, they don't feel anything. (Ohly, 2019, p. 17, translated by the author)

From this point of departure, one needs to take a further step and ask what exactly it is that robots replace. Robots neither replace humans as a whole nor their abilities, but they fulfil a task or a function that used to be done by humans. This means that they replace humans in fulfilling a task or a function. Consequently, "functional equivalences" are not suitable to grasp these technology-based possibilities when they are defined as follows:

> The same function can be fulfilled in different manners. Humans recognize the feeling of other humans by knowledge of human nature, intuition and empathy. Artificial intelligence can fulfil the same function by generating data – for example through face recognition systems that can recognize the mood of a person. Only what is being recognized is the same. (Ohly, 2019, p. 18)

Why can this notion of "functional equivalences" not be used to classify DS conceptually? When a DS fulfils a function or a task, it always remains just an approximation of the fulfilment of a function, or a task done by humans; it is not identical. For example, fulfilling the function of recognizing feelings in another human being is and remains different when it is done by a human being than if it is done by a machine, which gets close to the function of recognizing the feelings of another human being. It is similar to it but will never be the same because the machine – as elaborated above in Section 1.2 Data-Based Systems (DS) Rather Than "Artificial Intelligence" is lacking human emotionality. "Functional equivalences" remain under-complex and reductionist in the sense that fulfilling a function or a task never takes place in a vacuum, but in a context where humans, as subjects of fulfilling a function or a task, as well as the process of fulfilling a function or a task interact—either as a result of it or independently. For example, when humans fulfil the function of recognizing feelings in another human being, the humans are subjects of fulfilling this function, as well as the fulfilment of this function itself. The latter means that humans as subjects also change in the interaction with the observed humans, which affects the observed humans as well. And the feelings of the other human beings will change—as a result of it or independently.

Twenty-eighth, innovative methodological approaches fostering participation (Ruess et al., 2024), for example, "Citizen Development" (Binzer & Winkler, 2024), accelerate the technology-based transformation drastically by enabling everyone—even without any coding-know-how or low programming skills—to contribute substantially to the digital transformation of organizations and to the expansion of the use of DS, expanding significantly the digital transformation of human labor and the impact of DS on human work.

Twenty-ninth, the negative impact on paid professional tasks happens in an overall context of "low demand for labor" (Benanav, 2020) spinning faster an already rapid negative spiral further to the bottom of unemployment.

Thirtieth, this uniqueness of digital transformation and DS is extrapolated as fast as never before (because of effects of reciprocal reinforcement between investments and innovations) as well as constantly as never before (due to the race among researchers to publish as fast as possible any new tiny innovative detail out of a fear of missing the moment allowing someone else to publish first this new tiny innovative detail) as well as global as never before (because of a more globalized than ever economy) as transversal as never before, by a continuous amplification of the domains which are digitalized or where DS simulate the fulfillment of professional tasks so far in the hands of humans—for example, the current expansion by machine learning (Brynjolfsson et al., 2018) or by generative DS or by large language models (Adiasto, 2024; Barenkamp, 2023; Buxmann et al., 2024; Lips et al., 2023):

> In short, over the past decade, the potential scope of automation has expanded to include many virtual social interactions. The same is true of creative tasks that center on recombining existing ideas. Additionally, advancements in computer vision have paved the way for automating more perception tasks. (Frey & Osborne, 2024)

References

Acemoglu, D., & Restrepo, P. (2020). The wrong kind of AI? Artificial intelligence and the future of labour demand. *Cambridge Journal of Regions Economy and Society, 13*(1), 25–35. https://doi.org/10.1093/cjres/rsz022

Adiasto, K. (2024). Sustainable employability in the age of generative artificial Intelligence. *Group & Organization Management, 49*(6), 1338–1348. https://doi.org/10.1177/10596011241238792

Alimardani, A. (2024). Generative artificial intelligence vs. law students: An empirical study on criminal law exam performance. *Law Innovation and Technology, 16*(2), 777–819. https://doi.org/10.1080/17579961.2024.2392932

Alpsancar, S., Friedrich, A., Gehring, P., Kaminski, A., Hubig, C., & Nordmann, A. (2023). Editorial. In S. Alpsancar, A. Friedrich, P. Gehring, A. Kaminski, C. Hubig, & A. Nordmann (Eds.), *Faktor Mensch. Jahrbuch Technikphilosophie 2023* (9. Jahrgang) (pp. 9–14). Nomos.

Amankwah-Amoah, J., Abdalla, S., Mogaji, E., Elbanna, A., & Dwivedi, Y. K. (2024). The impending disruption of creative industries by generative AI: Opportunities, challenges, and research agenda. *International Journal of Information Management, 79*, 102759. https://doi.org/10.1016/j.ijinfomgt.2024.102759

Andrei, A. G., Maţcu-Zaharia, M., & Mariciuc, D. F. (2024). Ready to grip AI's potential? Insights from an exploratory study on perceptions of human-AI collaboration. *BRAIN: Broad Research in Artificial Intelligence & Neuroscience, 15*(2), 1–22. https://doi.org/10.18662/brain/15.2/560

Andrzejewski, C. (2023, February 15). 'Team Jorge': In the heart of a global disinformation machine. *Forbidden Stories.* Retrieved March 31, 2025, from https://forbiddenstories.org/story-killers/team-jorge-disinformation/

Atchley, P., Pannell, H., Wofford, K., Hopkins, M., & Atchley, R. A. (2024). Human and AI collaboration in the higher education environment: Opportunities and concerns. *Cognitive Research Principles and Implications, 9*(1). https://doi.org/10.1186/s41235-024-00547-9

Atuahene, B. T., Kanjanabootra, S., & Gajendran, T. (2022). Transformative role of big data through enabling capability recognition in construction. *Construction Management and Economics, 41*(3), 208–231. https://doi.org/10.1080/01446193.2022.2132523

Barbieri, L., Mussida, C., Piva, M., & Vivarelli, M. (2019). *Testing the employment impact of automation, robots and AI: A survey and some methodological issues* (IZA Discussion Paper No. 12612). IZA Institute of Labor Economics. https://doi.org/10.2139/ssrn.3457656

Barenkamp, M. (2023). Einflussnahme großer Sprachmodelle auf die moderne Arbeitswelt. *Informatik-Spektrum, 46*(4), 185–188. https://doi.org/10.1007/s00287-023-01546-8

Batista, J., Mesquita, A., & Carnaz, G. (2024). Generative AI and higher education: Trends, challenges, and future directions from a systematic literature review. *Information, 15*(11), 676. https://doi.org/10.3390/info15110676

Baumgartner, M., Kopp, T., Kick, E., Richter, D., & Kinkel, S. (2024). Travel counsellors' perspectives on AI assistance. *Zeitschrift Für Arbeitswissenschaft, 78*(3), 309–322. https://doi.org/10.1007/s41449-024-00431-8

Becker, M. (2017). *Automatisierung und Ausbeutung. Was wird aus der Arbeit im digitalen Kapitalismus?* Promedia.

Benanav, A. (2020). *Automation and the future of work.* Verso.

Bender, S. (2024). Generative-AI, the media industries, and the disappearance of human creative labour. *Media Practice and Education*, 1–18. https://doi.org/10.1080/25741136.2024.2355597

Bergstrom, T., & Ruediger, D. (2024, October 30). *A third transformation? Generative AI and scholarly publishing.* https://doi.org/10.18665/sr.321519

Binzer, B., & Winkler, T. J. (2024). Die vier Phasen von Citizen Development-Initiativen: Treiber, Herausforderungen und Handlungsempfehlungen. *HMD Praxis Der Wirtschaftsinformatik, 61*(5), 1136–1158. https://doi.org/10.1365/s40702-024-01088-x

Bond, M., Khosravi, H., De Laat, M., Bergdahl, N., Negrea, V., Oxley, E., Pham, P., Chong, S. W., & Siemens, G. (2024). A meta systematic review of artificial intelligence in higher education: A call for increased ethics, collaboration, and rigour. *International Journal of Educational Technology in Higher Education, 21*(1). https://doi.org/10.1186/s41239-023-00436-z

Boussioux, L., Lane, J. N., Zhang, M., Jacimovic, V., & Lakhani, K. R. (2024). The crowdless future? Generative AI and creative problem-solving. *Organization Science, 35*(5), 1589–1607. https://doi.org/10.1287/orsc.2023.18430

Broecke, S. (2023). *Artificial intelligence and labour market matching* (OECD Social, Employment and Migration Working Papers, No. 284). OECD Publishing. https://doi.org/10.1787/2b440821-en

Brynjolfsson, E., & McAffee, A. (2011). *Race against the machine: How the digital revolution is accelerating innovation, driving productivity, and irreversibly transforming employment and the economy.* Digital Frontier Press.

Brynjolfsson, E., & Mitchell, T. (2017). What can machine learning do? Workforce implications. *Science, 358*(6370), 1530–1534. https://doi.org/10.1126/science.aap8062

Brynjolfsson, E., Mitchell, T., & Rock, D. (2018). What can machines learn and what does it mean for occupations and the economy? *AEA Papers and Proceedings, 108*, 43–47. https://doi.org/10.1257/pandp.20181019

Buxmann, P., Glauben, A., & Hendriks, P. (2024). The use of ChatGPT in companies: A case study on the redesign of service processes. *HMD Praxis Der Wirtschaftsinformatik, 61*(2), 436–448. https://doi.org/10.1365/s40702-024-01053-8

Cadwalladr, C. (2024, August 3). 'A polarisation engine': how social media has created a 'perfect storm' for UK's far-right riots. *The Guardian*. Retrieved March 31, 2025, from https://www.theguardian.com/media/article/2024/aug/03/a-polarisation-engine-how-social-media-has-created-a-perfect-storm-for-uks-far-right-riots

Capraro, V., Lentsch, A., Acemoglu, D., Akgun, S., Akhmedova, A., Bilancini, E., Bonnefon, J., Brañas-Garza, P., Butera, L., Douglas, K. M., Everett, J. A. C., Gigerenzer, G., Greenhow, C., Hashimoto, D. A., Holt-Lunstad, J., Jetten, J., Johnson, S., Kunz, W. H., Longoni, C., & Viale, R. (2024). The impact of generative artificial intelligence on socioeconomic inequalities and policy making. *PNAS Nexus, 3*(6) https://doi.org/10.1093/pnasnexus/pgae191

Carbonero, F., Ernst, E., & Weber, E. (2018). *Robots worldwide: The impact of automation on employment and trade* (Research Department Working Paper No. 36). International Labour Organization. https://doi.org/10.13140/RG.2.2.10507.13603

Chalmers, D., MacKenzie, N. G., & Carter, S. (2020). Artificial intelligence and entrepreneurship: Implications for venture creation in the Fourth Industrial Revolution. *Entrepreneurship Theory and Practice, 45*(5), 1028–1053. https://doi.org/10.1177/1042258720934581

Chan, C. K. Y., & Tsi, L. H. (2024). Will generative AI replace teachers in higher education? A study of teacher and student perceptions. *Studies in Educational Evaluation, 83*, 101395. https://doi.org/10.1016/j.stueduc.2024.101395

Chesterman, S. (2024). Good models borrow, great models steal: Intellectual property rights and generative AI. *Policy and Society, 43*(3), 345–360. https://doi.org/10.1093/polsoc/puae006

Clarke, S., & Whittlestone, J. (2022). A survey of the potential long-term impacts of AI. In V. Conitzer, J. Tasioulas, M. Scheutz, R. Calo, M. Mara, & A. Zimmermann (Eds.), *Proceedings of the 2022 AAAI/ACM conference on AI, ethics, and society* (pp. 192–202). ACM. https://doi.org/10.1145/3514094.3534131

Clusmann, J., Kolbinger, F. R., Muti, H. S., Carrero, Z. I., Eckardt, J., Laleh, N. G., Loeffler, C. M. L., Schwarzkopf, S., Unger, M., Veldhuizen, G. P., Wagner, S. J., & Kather, J. N. (2023). The future landscape of large language models in medicine. *Communications Medicine, 3*(1). https://doi.org/10.1038/s43856-023-00370-1

Coetzee, M., & Veldsman, D. (2022). The digital-era industrial/organisational psychologist: Employers' view of key service roles, skills and attributes. *SA Journal of Industrial Psychology, 48.* https://doi.org/10.4102/sajip.v48i0.1991

Compagnucci, F., Gentili, A., Valentini, E., & Gallegati, M. (2019). Robotization and labour dislocation in the manufacturing sectors of OECD countries: A panel VAR approach. *Applied Economics, 51*(57), 6127–6138. https://doi.org/10.1080/00036846.2019.1659499

Cook, S., Hagiu, A., & Wright, J. (2024, January 1). Turn generative AI from an existential threat into a competitive advantage. *Harvard Business Review*. Retrieved March 31, 2025, from https://hbr.org/2024/01/turn-generative-ai-from-an-existential-threat-into-a-competitive-advantage

Corvalán, J. G. (2018). Digital and intelligent public administration: Transformations in the era of artificial intelligence. *A&C – Revista De Direito Administrativoe & Constitucional, 18*(71), 55–87. https://doi.org/10.21056/aec.v18i71.857

Crowe, B., Shah, S., Teng, D., Ma, P. S., DeCamp, M., Rosenberg, E. I., Rodriguez, J. A., Collins, B. X., Huber, K., Karches, K., Zucker, S., Kim, E. J., Rotenstein, L., Rodman, A., Jones, D., Richman, I. B., Henry, T. L., Somlo, D., Pitts, S. I., & Mishuris, R. G. (2024). Recommendations for clinicians, technologists, and healthcare organizations on the use of generative artificial intelligence in medicine: A position statement from the Society of General Internal Medicine. *Journal of General Internal Medicine, 40*(3), 694–702. https://doi.org/10.1007/s11606-024-09102-0

Cui, Z., Demirer, M., Jaffe, S., Musolff, L., Peng, S., & Salz, T. (2024). *The effects of generative AI on high skilled work: Evidence from three field experiments with software developers.* SSRN. https://doi.org/10.2139/ssrn.4945566

Davenport, T., Guha, A., Grewal, D., & Bressgott, T. (2019). How artificial intelligence will change the future of marketing. *Journal of the Academy of Marketing Science, 48*(1), 24–42. https://doi.org/10.1007/s11747-019-00696-0

De Luca Picione, G. L., Diana, P., Ferrari, G., Fortini, L., & Trezza, D. (2024). IA Generativa nel welfare: un approccio basato sulla Sociologia Pubblica per una governance consapevole. *Cambio Rivista Sulle Trasformazioni Sociali, 13*(26), 117–139. https://doi.org/10.36253/cambio-15358

De Vries, G. J., Gentile, E., Miroudot, S., & Wacker, K. M. (2020). The rise of robots and the fall of routine jobs. *Labour Economics, 66*, 101885. https://doi.org/10.1016/j.labeco.2020.101885

Demirci, O., Hannane, J., & Zhu, X. (2024). *Who is AI replacing? The impact of generative AI on online freelancing platforms.* CESifo Working Paper No. 11276. SSRN. https://doi.org/10.2139/ssrn.4991774

Deng, T., Wan, G., & Ma, M. (2024). Impact of tourism companies digital transformation on employment: Some evidence from China. *Asia Pacific Journal of Tourism Research, 29*(2), 225–238. https://doi.org/10.1080/10941665.2024.2324180

Dixon, J., Hong, B., & Wu, L. (2021). The robot revolution: Managerial and employment consequences for firms. *Management Science, 67*(9), 5586–5605. https://doi.org/10.1287/mnsc.2020.3812

Duffy, M. (2018). Building sustainable jobs and supporting human potential in the care sector. In E. Paus (Ed.), *Confronting Dystopia: The new technological revolution and the future of work* (pp. 94–112). Cornell University Press. https://doi.org/10.7591/9781501719868

Eloundou, T., Manning, S., Mishkin, P., & Rock, D. (2023). GPTs are GPTs: An early look at the labor market impact potential of large language models. *arXiv (Cornell University)*. Preprint retrieved from https://doi.org/10.48550/arxiv.2303.10130

Erickson, K. (2024). AI and work in the creative industries: Digital continuity or discontinuity? *Creative Industries Journal*, 1–21. https://doi.org/10.1080/17510694.2024.2421135

Fear, K., & Gleber, C. (2023). Shaping the future of older adult care: ChatGPT, advanced AI, and the transformation of clinical practice. *JMIR Aging, 6*, e51776. https://doi.org/10.2196/51776

Ferioli, E. A. (2019). L'intelligenza artificiale nei servizi sociali e sanitari: una nuova sfida al ruolo delle istituzioni pubbliche nel welfare italiano? *BioLaw Journal – Rivista Di BioDiritto, 15*(1), 163–175. https://doi.org/10.15168/2284-4503-358

Floridi, L. (2015). *Die 4. Revolution: Wie die Infosphäre unser Leben verändert.* Suhrkamp.

Fogleman, B. M., Goldman, M., Holland, A. B., Dyess, G., & Patel, A. (2024). Charting tomorrow's healthcare: A traditional literature review for an artificial intelligence-driven future. *Cureus, 16*(4), e58032. https://doi.org/10.7759/cureus.58032

Ford, M. (2015). *Rise of the robots: Technology and the threat of a jobless future.* Basic Books.

Frey, C. B., & Osborne, M. (2024). Generative AI and the future of work: A reappraisal. *Brown Journal of World Affairs, 30*(1), 1–17. https://bjwa.brown.edu/30-1/generative-ai-and-the-future-of-work-a-reappraisal/

Fu, X., Bao, Q., Xie, H., & Fu, X. (2020). Diffusion of industrial robotics and inclusive growth: Labour market evidence from cross country data. *Journal of Business Research, 122*, 670–684. https://doi.org/10.1016/j.jbusres.2020.05.051

Ghimire, P., Kim, K., & Acharya, M. (2024). Opportunities and challenges of generative AI in construction industry: Focusing on adoption of text-based models. *Buildings, 14*(1), 220. https://doi.org/10.3390/buildings14010220

Goldfarb, A., & Tucker, C. (2019). Digital economics. *Journal of Economic Literature, 57*(1), 3–43. https://doi.org/10.1257/jel.20171452

Gray, A. (2017, January 16). These are the world's 10 biggest corporate giants. *World Economic Forum*. Retrieved March 31, 2025, from https://www.weforum.org/stories/2017/01/worlds-biggest-corporate-giants/

Grewal, D., Guha, A., Satornino, C. B., & Becker, M. (2024). The future of marketing and marketing education. *Journal of Marketing Education, 47*(1), 61–77. https://doi.org/10.1177/02734753241269838

Gross, G. (2014, October 6). *The future of artificial intelligence: Will computers take your job?*. PCworld. Retrieved March 31, 2025, from https://www.pcworld.com/article/435671/the-future-of-artificial-intelligence-will-computers-take-your-job.html

Hartmann, E. A., Hornbostel, L., Thielicke, D. T., & Wittpahl, V. (2017). *Wie sieht die Zukunft der Arbeit aus? Ergebnisbericht zur Umfrage 'Künstliche Intelligenz und die Zukunft der Arbeit'*. Institut für Innovation und Technik Berlin (iit). Retrieved March 31, 2025, from https://www.iit-berlin.de/publikation/zukunft-der-arbeit/

Hartmann, P. (2023). AI in the crafts: Opportunities and challenges. In I. Knappertsbusch & K. Gondlach (Eds.), *Work and AI 2030* (pp. 219–226). Springer. https://doi.org/10.1007/978-3-658-40232-7_25

Hermida, A. (2024). From automata to algorithms: A jobs-to-be-done approach to AI in journalism. *Estudios Sobre El Mensaje Periodístico, 30*(3), 671–675. https://doi.org/10.5209/emp.97746

Hessler, M. (2016). Zur Persistenz der Argumente im Automatisierungsdiskurs. *Aus Politik und Zeitgeschichte, 66*(18–19), 17–24. Retrieved March 31, 2025, from https://www.bpb.de/apuz/225690/zur-persistenz-der-argumente-im-automatisierungsdiskurs

Hirsch-Kreinsen, H., & Karacic, A. (2019). Einleitung. In H. Hirsch-Kreinsen & A. Karacic (Eds.), *Autonome Systeme und Arbeit: Perspektiven, Herausforderungen und Grenzen der Künstlichen Intelligenz in der Arbeitswelt* (pp. 9–24). transcript.

Hofstetter, Y. (2017). Emanzipiert euch! Menschwerdung im digitalen Zeitalter. In B. Spiess & N. Fabisch (Eds.), *CSR und neue Arbeitswelten: Perspektivwechsel in Zeiten von Nachhaltigkeit, Digitalisierung und Industrie 4.0* (pp. 73–90). Management-Reihe Corporate Social Responsibility. Springer.

Horchi, M., & Ait Soudane, J. (2024). L'intégration de l'Intelligence Artificielle dans les pratiques de Gestion des Ressources Humaines: Une revue de littérature. *International Journal of Accounting, Finance, Auditing, Management and Economics, 5*(8), 275–290. https://doi.org/10.5281/zenodo.13328560

Huber, M., Jauch, C., & Burmeister, K. (2023). Potenziale der KI für die Produktion. In I. Knappertsbusch & K. Gondlach (Eds.), *Arbeit und KI 2030* (pp. 177–186). Springer. https://doi.org/10.1007/978-3-658-40232-7_20

Humbsch, P., Horn, E., Bohm, K., & Gintrowicz, R. (2024). ChatGPT im Einsatz für 'technology-enhanced learning' in Anästhesiologie und Notfallmedizin und potenzielle klinische Anwendung von KISprachmodellen. *Die Anaesthesiologie, 73*(5), 324–335. https://doi.org/10.1007/s00101-024-01403-7

Islam, I., & Islam, M. N. (2024). Exploring the opportunities and challenges of ChatGPT in academia. *Discover Education, 3*(1). https://doi.org/10.1007/s44217-024-00114-w

Jede, A., Bensberg, F., & Klein, T. (2024). Blockchain-technology in supply chain: Management–potentials and lacks of competencies. *HMD Praxis Der Wirtschaftsinformatik, 61*(1), 266–283. https://doi.org/10.1365/s40702-023-01029-0

Jošt, G., Taneski, V., & Karakatič, S. (2024). The impact of large language models on programming education and student learning outcomes. *Applied Sciences, 14*(10), 4115. https://doi.org/10.3390/app14104115

Juca-Maldonado, F. (2023). El impacto de la inteligencia artificial en los trabajos académicos y de investigación. *Revista Metropolitana de Ciencias Aplicadas, 6*(S1), 289–296. https://doi.org/10.62452/8nww1k83

Kaashoek, J., Raghavan, M., & Horton, J. J. (2024). The impact of generative AI on labor market matching. *An MIT Exploration of Generative AI.* https://doi.org/10.21428/e4baedd9.d562223c

Kan, Z., Zhao, S., Chi, W., & Zhou, J. (2024). AI types and organizational implications in various task contexts: A quantitative review from China. *Asia Pacific Business Review*, 1–28. https://doi.org/10.1080/13602381.2024.2409804

Kanbach, D. K., Heiduk, L., Blueher, G., Schreiter, M., & Lahmann, A. (2024). The GenAI is out of the bottle: Generative artificial intelligence from a business model innovation perspective. *Review of Managerial Science, 18*(4), 1189–1220. https://doi.org/10.1007/s11846-023-00696-z

Kirchschlaeger, P. G. (2016, March 30). Digitalisierung und Robotisierung der Gesellschaft aus theologisch-ethischer Perspektive. *Feinschwarz.net.* Retrieved March 31, 2025, from https://www.feinschwarz.net/digitalisierung-und-robotisierung-der-gesellschaft-aus-theologisch-ethischer-perspektive/

Kirchschlaeger, P. G. (2017). Roboter und Ethik. *Aktuelle Juristische Praxis, 26*(2), 240–249.

Kirchschlaeger, P. G. (2024). An International Data-Based Systems Agency IDA: Striving for a peaceful, sustainable, and human rights-based future. *Philosophies, 9*(3), 73. https://doi.org/10.3390/philosophies9030073

Klenert, D., Fernández-Macías, E., & Antón, J. (2022). Do robots really destroy jobs? Evidence from Europe. *Economic and Industrial Democracy, 44*(1), 280–316. https://doi.org/10.1177/0143831x211068891

Knight, G., & Khan, Z. (2022). Effects of emerging technologies on international business. In H. Merchant (Ed.), *The new frontiers of international business. Contributions to management science* (pp. 217–235). Springer. https://doi.org/10.1007/978-3-031-06003-8_10

Koegel, H., Spindler, M., & Wasserbacher, H. (2023). Digital finance – The future of financial planning in companies. In I. Knappertsbusch & K. Gondlach (Eds.), *Work and AI 2030* (pp. 159–166). Springer. https://doi.org/10.1007/978-3-658-40232-7_18

Krusche, M., Callhoff, J., Knitza, J., & Ruffer, N. (2023). Diagnostic accuracy of a large language model in rheumatology: Comparison of physician and ChatGPT-4. *Rheumatology International, 44*(2), 303–306. https://doi.org/10.1007/s00296-023-05464-6

Kuziemski, M., & Misuraca, G. (2020). AI governance in the public sector: Three tales from the frontiers of automated decision-making in democratic settings. *Telecommunications Policy, 44*(6), 101976. https://doi.org/10.1016/j.telpol.2020.101976

Langford, B. J., Branch-Elliman, W., Nori, P., Marra, A. R., & Bearman, G. (2024). Confronting the disruption of the infectious diseases workforce by artificial intelligence: What this means for us and what we can do about it. *Open Forum Infectious Diseases, 11*(3). https://doi.org/10.1093/ofid/ofae053

Larsson, S., White, J. M., & Bogusz, C. I. (2024). The artificial recruiter: Risks of discrimination in employers' use of AI and automated decision-making. *Social Inclusion, 12*(3), 1–18. https://doi.org/10.17645/si.7471

Lee, H. (2024). Reflecting on cultural labour in the time of AI. *Media Culture & Society, 46*(6), 1312–1323. https://doi.org/10.1177/01634437241254320

Levy, F., & Murnane, R. (2004). *The new division of labor.* Princeton University Press.

Leyh, C., & Schaeffer, T. (2024). Digitale Kompetenzen. *HMD Praxis Der Wirtschaftsinformatik, 61*(1), 1–5. https://doi.org/10.1365/s40702-024-01043-w

Li, Y., Choi, D., Chung, J., Kushman, N., Schrittwieser, J., Leblond, R., Eccles, T., Keeling, J., Gimeno, F., Lago, A. D., Hubert, T., Choy, P., De Masson D'Autume, C., Babuschkin, I., Chen, X., Huang, P., Welbl, J., Gowal, S., Cherepanov, A., & Vinyals, O. (2022). Competition-level code generation with AlphaCode. *Science, 378*(6624), 1092–1097. https://doi.org/10.1126/science.abq1158

Lips, M., Sahin, F., & Rosenauer, P. (2023). Effizienzsteigerungsmöglichkeiten durch Generative AI für Rechtsdienstleistungen. *Anwaltsrevue/Revue de l'avocat, 8*, 323–327. Retrieved March 31, 2025, from https://anwaltsrevue.recht.ch/de/artikel/02arv0823the/effizienzsteigerungsmoglichkeiten-durch-generative-ai-fur

Lopezosa, C., Pérez-Montoro, M., & Martín, C. R. (2024). El uso de la inteligencia artificial en las redacciones: propuestas y limitaciones. *Revista De Comunicación, 23*(1), 279–293. https://doi.org/10.26441/rc23.1-2024-3309

Martinelli, A., Mina, A., & Moggi, M. (2021). The enabling technologies of industry 4.0: Examining the seeds of the Fourth Industrial Revolution. *Industrial and Corporate Change, 30*(1), 161–188. https://doi.org/10.1093/icc/dtaa060

McKinsey & Company. (2024, October 17). *From job placement to disaster relief, McKinsey is using the power of AI for social good.* McKinsey Blog. Retrieved March 31, 2025, from https://www.mckinsey.com/about-us/new-at-mckinsey-blog/from-job-placement-to-disaster-relief-mckinsey-is-using-the-power-of-ai-for-social-good

Metzler, M. (2017, February 11). Martin Ford: "Automatisierung wird die ganze Arbeitswelt erfassen." *Neue Zuercher Zeitung.* https://www.nzz.ch/wirtschaft/martin-ford-automatisierung-wird-die-ganze-arbeitswelt-erfassen-ld.1802889

Mohammed, M. Y., & Skibniewski, M. J. (2023). The role of generative AI in managing industry Projects: Transforming industry 4.0 into industry 5.0 driven economy. *Law and Business, 3*(1), 27–41. https://doi.org/10.2478/law-2023-0006

Montejo, L., Fenton, A., & Davis, G. (2024). Artificial intelligence (AI) applications in healthcare and considerations for nursing education. *Nurse Education in Practice, 80*, 104158. https://doi.org/10.1016/j.nepr.2024.104158

Mueller, F., Sautner, P., Koemm, H., Vetter, G., & Zachner, N. (2024). *Quarterly executive survey on AI. KI-Studie: Die vier spannendsten Trends. Wenn es 2023 um Experimente ging, geht es 2024 um Ergebnisse.* Bain & Company. Retrieved March 31, 2025, from https://www.bain.com/de/insights/ai-umfrage/

Narayanan, A. [@random_walker]. (2025, Februar 26). One of the weirder scholarly practices regarding generative AI that seems to have been normalized is citing chatbots. [Tweet]. X. Retrieved March 31, 2025, from https://x.com/random_walker/status/1863664060119781779

Neuburger, R., & Fiedler, M. (2020). Zukunft der Arbeit – Implikationen und Herausforderungen durch autonome Informationssysteme. *Schmalenbach Journal of Business Research, 72*(3), 343–369. https://doi.org/10.1007/s41471-020-00097-y

Nordlinger, B., Kirchner, C., & De Fresnoye, O. (2024). Rapport 24-03. Systèmes d'IA générative en santé : enjeux et perspectives. *Bulletin De L Académie Nationale De Médecine, 208*(5), 536–547. https://doi.org/10.1016/j.banm.2024.03.005

Nowrozy, R. (2024). GPTs or grim position threats? The potential impacts of large language models on non-managerial jobs and certifications in cybersecurity. *Informatics, 11*(3), 45. https://doi.org/10.3390/informatics11030045

Noy, S., & Zhang, W. (2023). Experimental evidence on the productivity effects of generative artificial intelligence. *Science, 381*(6654), 187–192. https://doi.org/10.1126/science.adh2586

OECD. (2023). *Security risks in artificial intelligence.* OECD Global Forum on Digital Security for Prosperity, Paris. Retrieved March 31, 2025, from https://www.oecd.org/digital/global-forum-digital-security/GFDSP-2023-agenda.pdf

Ohly, L. (2019). *Ethik der Robotik und der Künstlichen Intelligenz. Theologisch-Philosophische Beiträge zu Gegenwartsfragen 22.* Peter Lang.

Ooi, K. B., Tan, G. W. H., Al-Emran, M., Al-Sharafi, M. A., Capatina, A., Chakraborty, A., & Wong, L. W. (2023). The potential of generative artificial intelligence across disciplines: Perspectives and future directions. *Journal of Computer Information Systems, 65*(1), 76–107. https://doi.org/10.1080/08874417.2023.2261010

Orchard, T., & Tasiemski, L. (2023). The rise of generative AI and possible effects on the economy. *Economics and Business Review/the Poznań University of Economics Review, 9*(2) https://doi.org/10.18559/ebr.2023.2.732

Osadchaya, E., Marder, B., Yule, J. A., Yau, A., Lavertu, L., Stylos, N., Oliver, S., Angell, R., De Regt, A., Gao, L., Qi, K., Zhang, W. Z., Zhang, Y., Li, J., & AlRabiah, S. (2024). To ChatGPT, or not to ChatGPT: Navigating the paradoxes of generative AI in the advertising industry. *Business Horizons, 67*(5), 571–581. https://doi.org/10.1016/j.bushor.2024.05.002

Owen-Smith, J. (2024, October 21). *AI and work: How to build new data foundations.* American Enterprise Institute. Retrieved March 31, 2025, from https://www.aei.org/research-products/report/ai-and-work-how-to-build-new-data-foundations/

Pelizzari, A., Pons-Vignon, N., Martinelli, A., & d'État à l'économie, S. (2024). *Les inspections du travail face aux transformations du marché du travail: une revue de la littérature Rapport final.* SECO. Retrieved March 31, 2025, from https://www.hetsl.ch/laress/catalogue-des-recherches/detail/recherche/les-inspections-du-travail-face-aux-transformations-du-marche-du-travail-ps82686

Pieterson, W., Baptista, D., Rosas-Shady, D., & Franco, A. (2023). *The digital transformation of public employment services across Latin America and the Caribbean.* Inter-American Development Bank. https://doi.org/10.18235/0005084

Poisat, P., Cullen, M., & Calitz, A. P. (2024). Human resource managers' perceptions on the impact of AI on the South African workforce. *SA Journal of Human Resource Management, 22*(0), a2593. https://doi.org/10.4102/sajhrm.v22i0.2593

Prasad, K. D. V., & De, T. (2024). Generative AI as a catalyst for HRM practices: Mediating effects of trust. *Humanities and Social Sciences Communications, 11*(1). https://doi.org/10.1057/s41599-024-03842-4

Pressman, S. M., Borna, S., Gomez-Cabello, C. A., Haider, S. A., Haider, C. R., & Forte, A. J. (2024). Clinical and surgical applications of large language models: A systematic review. *Journal of Clinical Medicine, 13*(11), 3041. https://doi.org/10.3390/jcm13113041

Raj, M., & Seamans, R. (2019). Primer on artificial intelligence and robotics. *Journal of Organization Design, 8*(1). https://doi.org/10.1186/s41469-019-0050-0

Ransbotham, S., Kiron, D., Khodabandeh, S., Chu, M., & Zhukov L. (2024, November 12). Learning to manage uncertainty, with AI. *MIT Sloan Management Review and Boston Consulting Group*. Retrieved March 31, 2025, from https://www.bcg.com/publications/2024/manage-uncertainty-with-organizational-learning-ai

Reddy, S. (2024). Generative AI in healthcare: An implementation science informed translational path on application, integration and governance. *Implementation Science, 19*(1). https://doi.org/10.1186/s13012-024-01357-9

Romera-Paredes, B., Barekatain, M., Novikov, A., Balog, M., Kumar, M. P., Dupont, E., Ruiz, F. J. R., Ellenberg, J. S., Wang, P., Fawzi, O., Kohli, P., & Fawzi, A. (2023). Mathematical discoveries from program search with large language models. *Nature, 625*(7995), 468–475. https://doi.org/10.1038/s41586-023-06924-6

Rotman, D. (2013). How technology is destroying jobs. *MIT Technology Review, 16*(*4*), *28–35.* Retrieved March 31, 2025, from https://www.technologyreview.com/s/515926/how-technology-is-destroying-jobs/

Ruediger, D., McCracken, C., & Skinner, M. (2024). *Adoption of generative AI by academic biomedical researchers.* ITHAKA S+R. http://www.jstor.org/stable/resrep64145

Ruehr, A., Berger, B., & Hess, T. (2023). Intelligent IT systems in business application. In I. Knappertsbusch & K. Gondlach (Eds.), *Work and AI 2030* (pp. 125–132) https://doi.org/10.1007/978-3-658-40232-7_14

Ruess, P., Staffa, A., Kreutz, A., Busch, C., Gayoso, C. O. S., & Pollmann, K. (2024). Künstliche Intelligenz in betrieblichen Prozessen: Ein Vorgehensmodell zur partizipativen Gestaltung von KI-Anwendungen. *HMD Praxis Der Wirtschaftsinformatik, 61*(2), 485–502. https://doi.org/10.1365/s40702-024-01049-4

Ruh, N., Meissner, J. O., Burri, D., Camp, M.-A., Frei, S., Gretler Heusser, S., Grueter, U., Hauser, C., Huebscher, B., Klotz, U., Murri, M., Odermatt, I., & Wey, Y. (2024). Einfluss der Digitalisierung auf die Kulturbereiche Musik, Theater und Visuelles Design. In TA-SWISS Publikationsreihe (Hrsg.), *Kultur und Digitalisierung* (TA 83/2024, pp. 25–289). vdf.

Russell, S. (2021, December 22). Living with artificial intelligence, *The Reith Lectures* [Radio feature]. BBC Radio 4. https://www.bbc.co.uk/programmes/m001216k

Sako, M. (2024). How generative AI fits into knowledge work. *Communications of the ACM, 67*(4), 20–22. https://doi.org/10.1145/3638567

Sandel, M. (2007). *The case against perfection: Ethics in the age of genetic engineering*. Harvard University Press.

Schmidt, D. A. (2023). AI in banks: The bank of the future. In I. Knappertsbusch & K. Gondlach (Eds.), *Work and AI 2030* (pp. 167–173). Springer. https://doi.org/10.1007/978-3-658-40232-7_19

Schoenberger, M., & Beinke, J. H. (2023). Hybride Intelligenz als Konvergenz menschlicher und künstlicher Intelligenz – wie verändert ChatGPT die Wissensarbeit? *HMD Praxis Der Wirtschaftsinformatik, 60*(6), 1174–1193. https://doi.org/10.1365/s40702-023-00997-7

Schwab, K. (2016). *The fourth industrial revolution*. Crown Business.

Séïde, M. (2024). Éducation et intelligence artificielle générative: Contribution de l'UNESCO à partir des publications récentes. *Rivista di Scienze dell'Educazione, 62*(1), 54–70. Retrieved March 31, 2025, from https://ethik-und-gesellschaft.de/ojs/index.php/rse/article/view/2004/2067

Seufert, S., & Meier, C. (2023). Hybrid Intelligence: Collaboration with AI systems for knowledge work. *HMD Praxis Der Wirtschaftsinformatik, 60*(6), 1194–1209. https://doi.org/10.1365/s40702-023-01012-9

Smith, N. (2013, January 14). The end of labor: How to protect workers from the rise of robots. *The Atlantic*. Retrieved March 31, 2025, from https://www.theatlantic.com/business/archive/2013/01/the-end-of-labor-how-to-protect-workers-from-the-rise-of-robots/267135/

Stark, D., & Pais, I. (2020). Algorithmic management in the platform economy. *Sociologica, 14*(3), 47–72. https://doi.org/10.6092/issn.1971-8853/12221

Stowasser, S. (2023). Successful introduction of AI in the company. In I. Knappertsbusch & K. Gondlach (Eds.), *Work and AI 2030* (pp. 133–141). Springer. https://doi.org/10.1007/978-3-658-40232-7_15

Susskind, R. E., & Susskind, D. (2015). *The future of the professions: How technology will transform the work of human experts*. Oxford University Press.

Taddy, M. (2018). *The technological elements of artificial intelligence* (NBER Working Paper 243001). National Bureau of Economic Research. https://doi.org/10.3386/w24301

Varnosfaderani, S. M., & Forouzanfar, M. (2024). The role of AI in hospitals and clinics: Transforming healthcare in the 21st century. *Bioengineering, 11*(4), 337. https://doi.org/10.3390/bioengineering11040337

Verma, P., & Zakrzewski, C. (2024, April 23). AI deepfakes threaten to upend global elections. No one can stop them. *The Washington Post*. Retrieved March 31, 2025, from https://www.washingtonpost.com/technology/2024/04/23/ai-deepfake-election-2024-us-india/

Wachter, R. M., & Brynjolfsson, E. (2023). Will generative artificial intelligence deliver on its promise in health care? *JAMA, 331*(1), 65. https://doi.org/10.1001/jama.2023.25054

Watermeyer, R., Phipps, L., Lanclos, D., & Knight, C. (2023). Generative AI and the automating of academia. *Postdigital Science and Education, 6*(2), 446–466. https://doi.org/10.1007/s42438-023-00440-6

Wenz, F., & Ebener, S. (2024). Anwendungen künstlicher Intelligenz in der Onkologie: Möglichkeiten, Machbarkeit und regulatorische Herausforderungen. *Onkologie, 30*(5), 339–346. https://doi.org/10.1007/s00761-023-01428-4

Wilhelms, G. (2018). Wie die Digitalisierung das Verhältnis des Menschen zu seiner Arbeit verändert: Versuch einer Kritik. *AMOS. International, 12*(1), 3. Retrieved March 31, 2025, from https://www.amosinternational.de/magazine/issue-2018-1

Wilson, J. H., & Daugherty, P. R. (2018). Collaborative intelligence: Humans and AI are joining forces. *Harvard Business Review, 96*(4), 114–123.

Xia, Q., Weng, X., Ouyang, F., Lin, T. J., & Chiu, T. K. (2024). A scoping review on how generative artificial intelligence transforms assessment in higher education. *International Journal of Educational Technology in Higher Education, 21*, 40. https://doi.org/10.1186/s41239-024-00468-z

Zuboff, S. (2019). *The age of surveillance capitalism: The fight for a human future at the new frontier of Power*. PublicAffairs.

CHAPTER 4

Massive Reduction of Paid Professional Tasks

Abstract In this chapter, the massive reduction of paid professional tasks for humans is ethically assessed. From an ethical point of view, fewer working hours and more free time are not necessarily a bad thing. It becomes ethically problematic if the disappearing sources of income, of financial means for social welfare, but also of other aspects of paid professional tasks like providing purpose of life or identity for humans are not addressed and if the core consequence of digital transformation and the use of data-based systems (DS) is not considered: Fewer and fewer humans will directly participate and partake in a more efficient and effective value chain.

Keywords Massive reduction of paid professional tasks for humans • Ethical assessment • Income • Social welfare • Functions of paid professional tasks • Purpose of life • Identity • Value chain

These *30 differentiators of digital transformation and the use of DS from earlier technology-based eras of change* elaborated above undermine the main argument against the massive reduction of paid professional tasks for humans, namely that never in human history have innovation-based shifts led to a decrease in jobs. This way, these 30 differentiators of digital transformation and the use of DS from earlier technology-based eras of change

P. G. Kirchschlaeger, *Ethics and the Digital Transformation of Human Work*, https://doi.org/10.1007/978-3-031-98704-5_4

form a first argument favoring a massive reduction of paid professional tasks for humans.

A second argument supporting a massive reduction of paid professional tasks by digital transformation and DS also justifies the focus on "paid professional tasks" (Agrawal et al., 2019) rather than on paid jobs (Bonney et al., 2024a).

> There are two main problems with how AI and automation are discussed today. The first is a sense that they are somehow new and unprecedented interventions into the labor process. Historically, the machine is an attempt to automate part of the labor process, increasing the amount produced by a worker. It has a significantly longer history than self-driving cars or automated warehouse pickers. This leads us to the second problem: there is often a binary understanding of automation, either something will be automated or not. This leads to a focus on the machine – how effective is the new self-driving vehicle, for example – rather than looking at how automation is actually affecting work and workers. Rather than an either / or, automation is used much more as an augmentation of the labor process. (Moore & Woodcock, 2021, p. 1)

While it has been attempted to address the first problem by examining the relationship between the digital transformation and the use of DS and previous transformational periods caused by innovation above, the second problem provoked not only to analyze more precisely single professional tasks rather than entire jobs but also to look more closely at the impact of the digital transformation and the use of DS on the working humans in a holistic way which results in the finding that working humans are to be replaced in the course of digital transformation and the use of DS.

A third argument encompasses several studies demonstrating that there will be a disruption (Panzaru & Negoiță, 2022, p. 28) (this has already occurred with, e.g., travel industry and online-booking; music industry and streaming; retail-industry and e-shopping; etc. (Meyer, 2015)) and a massive reduction of paid professional tasks (Acemoglu, 2023; Acemoglu et al., 2022; Acemoglu & Johnson, 2024; Acemoglu & Restrepo, 2019b; Aguilera & Barrera, 2016; Autor & Salomons, 2018; Ayad & El Mezouari, 2024; Barbosa et al., 2022; Bell & Korinek, 2023; Bender, 2024; Benzell et al., 2015; Bonney et al., 2024a; Bonney et al., 2024b; Carbonero et al., 2018; Compagnucci et al., 2019; Cusumano et al., 2024; De Vries et al., 2020; Demirci et al., 2024; Dengler & Matthes, 2018; Eloundou et al., 2024; Felten et al., 2024; Foladori & Ortiz-Espinoza, 2022; Freeman,

2015; Frey & Osborne, 2017; Hui et al., 2024; Manyika et al., 2017; McFee, 2024; McKinsey Global Institute, 2017; Paus, 2018; Pirvu et al., 2024; Prettner et al., 2018; Restrepo, 2023; Rueda & Thewissen, 2018; Webb, 2019; Wong, 2024; World Bank, 2016; Wu & Deng, 2024), e.g., that today's general-purpose DS could affect 60% of current jobs in advanced economies (Cazzaniga et al., 2024) or that half of today's work activities could be automated by 2055 (Ateneo Impresa, n.d.). Or as it is framed in another study consisting of a large survey of machine learning researchers on AI advancement already *in 2018*:

> AI will outperform humans in many activities in the next ten years, such as translating languages (by 2024), writing high-school essays (by 2026), driving a truck (by 2027), working in retail (by 2031), writing a bestselling book (by 2049), and working as a surgeon (by 2053). Researchers believe there is a 50% chance of AI outperforming humans in all tasks in 45 years and of automating all human jobs in 120 years, with Asian respondents expecting these dates much sooner than North Americans. (Grace et al., 2018, p. 729)

Based on the status quo of technological progress already in 2018, the McKinsey Global Institute assessed: "Even though AI and automation bring clear benefits to businesses and the economy by raising productivity growth and innovation, major disruptions to work can be expected, about half of current work activities (not jobs) being technically automatable" (Manyika & Bughin, 2018, p. 4). The reduction in paid professional tasks for humans and, thus, in paid jobs is proving to be an effect of digital transformation and DS, and in particular, of the use of self-learning systems which took place in the process, and poses great challenges—including abstraction in the sense of a distance from humans, their freedom, and their will to create, as well as immediacy as a description of the phenomenon of humans being put under the spell of machines (Wilhelms, 2018). Sebastian Siemiatkowski, chief executive and a co-founder of the technology-firm Klarna, says in 2025: "I am of the opinion that A.I. can already do all of the jobs that we, as humans, do. [...] Not only can we do more with less, but we can do much more with less" (Siemiatkowski, as cited in Scheiber, 2025).

A fourth argument is "that if the origin of productivity growth in the future continues to be automation, the relative standing of labor, together with the task content of production, will decline" (Acemoglu & Restrepo,

2019a, p. 27). As one of the characteristics of this technology-based transformation consists in creating self-learning machines which depend in their further development on small or no input by humans and, this way, serve the striving for efficiency and for productivity growth (Damioli et al., 2021), it is plausible to expect automation to remain the origin of productivity growth. Given this economic focus on automation consisting of replacing humans by DS, the consequence will be a massive reduction of paid professional tasks for humans.

A fifth argument represents the "paradox of potent impotence" (Kirchschlaeger, 2021). Like never before in the history of humanity, humans possess the power to create technology-based realities and/or a reality with a technological penetration, imprint, and dominance never seen before. DS always play a role; DS are everywhere; everything is data-based. "Technics, as a universe of instrumentalities, may increase the weakness as well as the power of man" (Marcuse, 1964, p. 185). In other words, present-day humans have higher potency in creating reality than ever before. An unprecedented portion of the world is being created by humans. At the same time, due to the enormous pace of technological progress in the domain of digital transformation and DS, due to the intransparency (Doshi-Velez et al., 2017; Kosack & Fung, 2014; Wachter et al., 2017) of some processes, due to the untraceability of some results because of the intransparency of some processes, and due to the high complexity of digital transformation and DS, humans are impotent at foreseeing their actions and the consequences of their actions.

> The gap between the ability to foretell and the power to act creates a novel moral problem. With the latter so superior to the former, recognition of ignorance becomes the obverse of the duty to know and thus part of the ethics that must govern the evermore necessary self-policing of our outsized might. (Jonas, 1985, p. 28, translated by the author)

This impotence is extended further from the perspective of the individual. "The I, the free subjectivity, has created these factual orders; but it no longer knows how to encompass them and no longer knows how to penetrate them with itself" (Cassirer, 1985, p. 76; Ellul, 1964, p. 418). As the collective forces itself into the foreground as the acting, the thinking and acting individual is rolled over, pushed aside, or forced into the background.

> If nothing succeeds like success, nothing captures like success. Whatever else belongs to the fullness of man is outshined in prestige by the expansion of his power, and so this expansion is, by binding more and more of man's powers to its business, accompanied by a shrinking of his self-concept and being. In the image that he maintains of himself – the programmatic conception that determines his current being as much as it reflects it – man is now more and more the maker of what he [or she] has produced and the door of what he [or she] will soon be able to do. Who is 'he' [or 'she']? Not you or me: it is the collective actor and the collective act, not the individual actor and the individual act that play a role here; it is the indefinite future much more than the contemporary space of action that provides the relevant horizon of responsibility. (Jonas, 1985, p. 32, translated by the author)

Beyond that, humans perceive themselves as powerless while trying legitimately and necessarily to identify the dimensions, relations, and spheres of responsibility—e.g., with the idea of an "ethical black box" (Winfield & Jirotka, 2017) in DS.

> A core experience in our contemporary socio-technical lifeworld – often resulting in fear – concerns responsibility and accountability: namely, the difficulty to attribute responsibility and to locate accountability in ever more distributed and entangled socio-technical systems. Think small: about the difficulties of finding and reaching the person to make responsible in case of a non-functioning internet connection? Think big: who's responsible – accountable and liable – for the financial crisis? (Simon, 2015, p. 136)

Humans are trapped in this paradox of potent impotence, which affects the ethical discussion of digital transformation of human work and the impact of the use of DS on human labor. The advantage of ***homo faber socialis*** over ***homo sapiens singularis***, and the impression that it currently looks like a victory for ***homo faber socialis***, challenges ethics fundamentally due to the significance of the autonomy of the individual for human dignity and for ethics as such.

> Each time we thus bypass the human way of dealing with human problems, short-circuiting it by an impersonal mechanism, we have taken away something from the dignity of personal selfhood and advanced one step further on the road from responsible subjects to programmed behavior systems. (Jonas, 1985, p. 52, translated by the author)

Could DS represent these systems? *Homo faber singularis* does not give up, though. "Technology, apparatus and mass existence do not exhaust human existence. [...] They come upon himself [herself], who is something else" (Jaspers, 1931, p. 61).

As this paradox of potent impotence consists of humans excluding powerfully humans from power—power embracing also paid professional tasks—it represents another argument favoring the massive reduction of paid professional tasks for humans.

The impact of the paradox of potent impotence on the substitution of humans by DS on the labor market is enhanced by an even more fundamental phenomenon, namely that the paradox of potent impotence expresses itself in the striving for DS *per se*:

> If men create intelligent machines, or fantasize about them, it is either because they secretly despair for their own intelligence or because they are in danger of succumbing to the weight of a monstrous and useless intelligence which they seek to exorcise by transferring it to machines, where they can play with it and make fun of it. By entrusting this burdensome intelligence to machines, we are released from any responsibility to knowledge, much as entrusting power to politicians allows us to disdain any aspiration of our own to power. [...] Just as eyeglasses and contact lenses will arguably evolve one day into implanted prostheses for a species that has lost sight, it is similarly to be feared that artificial intelligence and the hardware that supports it will become a mental prosthesis for a species without the capacity for thought. (Baudrillard, 1993, pp. 51–52)

Humans run the risk of taking themselves out of the game, stealing the wheel, and leaving it to DS (Kaeser, 2019; Spielkamp, 2019). Potent humans run the risk of making themselves impotent. Potent humans run the risk of substituting themselves on the labor market by DS making themselves impotent.

A sixth argument consists of the "empowerment of DS and human fear of powerlessness" (Kirchschlaeger, 2021): On the one hand, humans invest promising talent and creativity as well as enormous financial means, resources, energy, time, and research-focus to develop and create technologies that become more intelligent, more efficient, more effective, smarter, faster, and better than humans across a variety of domains (instead of prioritizing, e.g., pressing current global issues like global poverty, climate destruction, etc.). DS should evolve from tools to assistants, to teammates, to companions, to coaches, and to bosses. DS should obtain omnipresence (including ubiquitous DS), omniscience (including "super-DS"), and omnipotence (including, e.g., creativity in the full sense of the word). On the other hand, humans feel threatened by, and want to avoid, the possible future where this more intelligent, more efficient, more effective, smarter, faster, and better technologies take over control. Although, it perhaps could even make sense that—in certain domains—the power held by technologies is more intelligent, smarter, and better than it would be in the hands of human, humans absolutely want to remain in charge.

This paradox has its roots in the fear of humans of being disenthralled by technological systems (Neuhaeuser, 2014, p. 269), with ethically justifiable trust in humans, and with ethically justifiable non-existent willingness to entrust leadership to technologies due to the unique moral capability of humans and the lack of moral capability of technologies as elaborated above in Subchapter 1.2 "Data-Based Systems (DS) Rather Than 'Artificial Intelligence'" (Kirchschlaeger, 2017).

Even if DS can be encoded and trained with ethical principles as well as ethical and legal norms, which they must respect and implement, the high probability that self-learning DS in either the near or distant future will become "super-DS," and thereby outperform humans and human intelligence, results in the potential of a future reality wherein "super-DS" could also decide not to follow either ethical principles or ethical and legal norms anymore causing inhumane and unsustainable consequences for humanity and the planet.

Even the possibility that they could auto-define what "ethical" and "legal" should mean (although maybe it would not be "ethical" or "legal" at all because, e.g., it would serve only their self-interests and would not respect human dignity of all humans or the planet) must be considered from an ethical perspective. The question arises of what will happen to humans if DS take over power—power in a holistic sense including, e.g., cognitive power—and humans do not have anything to say anymore. What can be done today to avoid that in the future DS outpace humans and, in power, treat humans badly and do not respect the human dignity of all humans? What can be done today to prevent future DS from treating humans in the same way or worse than humans treat animals today? What can be done today that DS are encoded and trained to still respect humans as their masters although they outperform humans (Manzeschke & Karsch, 2016, p. 10), to uphold and to realize human rights, as well as to ensure that the universality and the validity of human rights are never open to change by DS? What can be done today to avoid that DS treat humans like slaves? What can be done today to set a clear priority on the "usefulness to life" (Justitia et Pax, 1998, pp. 22–26) instead of feasibility? What can be done today to stop and to avoid the substitution of humans by DS on the labor market?

The empowerment of DS and human fear of powerlessness forms another argument showing that many more paid professional tasks will disappear than will be created because of digital transformation and DS (Betschon, 2015; Brynjolfsson & McAffee, 2011; Cowen, 2013; Ford, 2009, 2015; Frey & Osborne, 2017; Manyika et al., 2017; Ohly, 2019, pp. 126–133; Schweighofer, 2016), which significantly distinguishes this transformation from previous boosts of innovation.

References

Acemoglu, D. (2023). Harms of AI. In J. B. Bullock (Ed.), *The Oxford handbook of AI governance* (pp. 660–706). Oxford University Press. https://doi.org/10.1093/oxfordhb/9780197579329.013.65

Acemoglu, D., Anderson, G., Beede, D., Buffington, C., Childress, E., Dinlersoz, E., Foster, L., Goldschlag, N., Haltiwanger, J., Kroff, Z., Restrepo, P., & Zolas, N. (2022). *Automation and the workforce: A firm-level view from the 2019 Annual Business Survey* (NBER Working Paper No. 30659). National Bureau of Economic Research. https://doi.org/10.2139/ssrn.4282509

Acemoglu, D., & Johnson, S. (2024). Learning from Ricardo and Thompson: Machinery and labor in the early Industrial Revolution and in the age of artificial intelligence. *Annual Review of Economics, 16*(1), 597–621. https://doi.org/10.1146/annurev-economics-091823-025129

Acemoglu, D., & Restrepo, P. (2019a). Automation and new tasks: How technology displaces and reinstates labor. *The Journal of Economic Perspectives, 33*(2), 3–30. https://doi.org/10.1257/jep.33.2.3

Acemoglu, D., & Restrepo, P. (2019b). The wrong kind of AI? Artificial intelligence and the future of labour demand. *Cambridge Journal of Regions Economy and Society, 13*(1), 25–35. https://doi.org/10.1093/cjres/rsz022

Agrawal, A., Gans, J. S., & Goldfarb, A. (2019). Artificial intelligence: The ambiguous labor market impact of automating prediction. *The Journal of Economic Perspectives, 33*(2), 31–50. https://doi.org/10.1257/jep.33.2.31

Aguilera, A., & Barrera, M. G. R. (2016). Technological Unemployment: An approximation to the Latin American Case. *AD-minister, 29*, 59–78. https://doi.org/10.17230/ad-minister.29.3

Ateneo Impresa. (n.d.). *Intelligenza artificiale e impatto sul mondo del lavoro.* Retrieved March 31, 2025, from https://www.ateneoimpresa.it/post/intelligenza-artificiale-e-impatto-sul-mondo-del-lavoro

Autor, D., & Salomons, A. (2018). *Is automation labor-displacing? Productivity growth, employment, and the labor share* (NBER Working Paper No 24871). National Bureau of Economic Research. https://doi.org/10.3386/w24871

Ayad, M., & El Mezouari, S. (2024). Investigating opportunities, challenges, and threats of artificial intelligence in accounting functions using a systematic literature review. *African Scientific Journal, 3*(24), 841. https://doi.org/10.5281/zenodo.12926948

Barbosa, C. E., De Lima, Y. O., Costa, L. F. C., Santos, H. S. D., Lyra, A., Argôlo, M., Da Silva, J. A., & De Souza, J. M. (2022). Future of work in 2050: Thinking beyond the COVID-19 pandemic. *European Journal of Futures Research, 10*, 25. https://doi.org/10.1186/s40309-022-00210-w

Baudrillard, J. (1993). Xeros and infinity. In J. Baudrillard (Ed.), *The transparency of evil: Essays on extreme phenomena* (pp. 51–59). Verso.

Bell, S. A., & Korinek, A. (2023). AI's economic peril. *Journal of Democracy, 34*(4), 151–161. https://doi.org/10.1353/jod.2023.a907696

Bender, S. (2024). Generative-AI, the media industries, and the disappearance of human creative labour. *Media Practice and Education, 25*(2), 1–18. https://doi.org/10.1080/25741136.2024.2355597

Benzell, S., Kotlikoff, L., LaGarda, G., & Sachs, J. (2015). *Robots are us: Some economics of human replacement* (NBER Working Paper No. 20941). National Bureau of Economic Research. https://doi.org/10.3386/w20941

Betschon, S. (2015, August 21). Automatisierung. Ein neues Maschinenzeitalter. *Neue Zuercher Zeitung*. Retrieved March 31, 2025, from https://www.nzz.ch/gesellschaft/lebensart/gesellschaft/ein-neues-maschinenzeitalter-1.18599135

Bonney, K., Breaux, C., Buffington, C., Dinlersoz, E., Foster, L., Goldschlag, N., Haltiwanger, J., Kroff, Z., & Savage, K. (2024a). The impact of AI on the workforce: Tasks versus jobs? *Economics Letters, 244*, 111971. https://doi.org/10.1016/j.econlet.2024.111971

Bonney, K., Breaux, C., Buffington, C., Dinlersoz, E., Foster, L., Goldschlag, N., Haltiwanger, J., Kroff, Z., & Savage, K. (2024b). *Tracking firm use of AI in real time: A snapshot from the business trends and outlook survey* (NBER Working Paper No. W32319). National Bureau of Economic Research. https://doi.org/10.2139/ssrn.4794368

Brynjolfsson, E., & McAffee, A. (2011). *Race against the machine: How the digital revolution is accelerating innovation, driving productivity, and irreversibly transforming employment and the economy*. Digital Frontier Press.

Carbonero, F., Ernst, E., & Weber, E. (2018). *Robots worldwide: The impact of automation on employment and trade* (Research Department Working Paper No. 36). International Labour Organization. https://doi.org/10.13140/RG.2.2.10507.13603

Cassirer, E. (1985). Form und Technik. In E. Cassirer (Ed.), *Symbol, Technik, Sprache* (pp. 39–91). Felix Meiner.

Cazzaniga, M., Jaumotte, F., Li, L., Melina, G., Panton, A. J., Pizzinelli, C., Rockall, E. J., & Tavares, M. M. (2024). *Gen-AI: Artificial intelligence and the future of work* (IMF Staff Discussion Note No. 2024/001). International Monetary Fund. https://doi.org/10.5089/9798400262548.006

Compagnucci, F., Gentili, A., Valentini, E., & Gallegati, M. (2019). Robotization and labour dislocation in the manufacturing sectors of OECD countries: A panel VAR approach. *Applied Economics, 51*(57), 6127–6138. https://doi.org/10.1080/00036846.2019.1659499

Cowen, T. (2013). *Average is over: Powering America beyond the age of the great stagnation*. Dutton.

Cusumano, M. A., Farias, V. F., & Ramakrishnan, R. (2024). Generative AI as a new platform for applications development. *An MIT Exploration of Generative AI*. https://doi.org/10.21428/e4baedd9.f189351f

Damioli, G., Van Roy, V., & Vertesy, D. (2021). The impact of artificial intelligence on labor productivity. *Eurasian Economic Review, 11*(1), 1–25. https://doi.org/10.1007/s40821-020-00172-8

De Vries, G. J., Gentile, E., Miroudot, S., & Wacker, K. M. (2020). The rise of robots and the fall of routine jobs. *Labour Economics, 66*, 101885. https://doi.org/10.1016/j.labeco.2020.101885

Demirci, O., Hannane, J., & Zhu, X. (2024). *Who is AI replacing? The impact of generative AI on online freelancing platforms* (CESifo Working Paper No. 11276). Social Science Research Network. https://doi.org/10.2139/ssrn.4991774

Dengler, K., & Matthes, B. (2018). The impacts of digital transformation on the labour market: Substitution potentials of occupations in Germany. *Technological Forecasting and Social Change, 137*, 304–316. https://doi.org/10.1016/j.techfore.2018.09.024

Doshi-Velez, F., Kortz, M., Budish, R., Bavitz, C., Gershman, S., O'Brien, D, Scott, K., Shieber, S., Waldo, J., Weinberger, D., Weller, A., & Wood, A. (2017). *Accountability of AI under the law: The role of explanation*. arXiv. Preprint Retrieved March 31, 2025, from https://arxiv.org/ftp/arxiv/papers/1711/1711.01134.pdf

Ellul, J. (1964). *The technological society*. Vintage Books.

Eloundou, T., Manning, S., Mishkin, P., & Rock, D. (2024). GPTs are GPTs: Labor market impact potential of LLMs. *Science, 384*(6702), 1306–1308. https://doi.org/10.1126/science.adj0998

Felten, E., Raj, M., & Seamans, R. (2024). Generative AI requires broad labor policy considerations. *Communications of the ACM, 67*(8), 29–32. https://doi.org/10.1145/3637864

Foladori, G., & Ortiz-Espinoza, Á. (2022). La relación capital-trabajo en la Industria 4.0. *Íconos – Revista De Ciencias Sociales, 26*(73), 161–177. https://doi.org/10.17141/iconos.73.2022.5198

Ford, M. (2009). *The lights in the tunnel: Automation, accelerating technology and the economy of the future*. Acculant Publishing.

Ford, M. (2015). *Rise of the robots: Technology and the threat of a jobless future*. Basic Books.

Freeman, R. (2015). Who owns the robots rules the world. *IZA World of Labor*. https://doi.org/10.15185/izawol.5

Frey, C. B., & Osborne, M. A. (2017). The future of employment: How susceptible are jobs to computerisation? *Technological Forecasting and Social Change, 114*, 254–280. https://doi.org/10.1016/j.techfore.2016.08.019

Grace, K., Salvatier, J., Dafoe, A., Zhang, B., & Evans, O. (2018). Viewpoint: When will AI exceed human performance? Evidence from AI experts. *Journal Of Artificial Intelligence Research, 62*, 729–754. https://doi.org/10.1613/jair.1.11222

Hui, X., Reshef, O., & Zhou, L. (2024). The short-term effects of generative artificial intelligence on employment: Evidence from an online labor market. *Organization Science, 35*(6), 1–20. https://doi.org/10.1287/orsc.2023.18441

Jaspers, K. (1931). *Die geistige Situation der Zeit*. De Gruyter.

Jonas, H. (1985). *Das Prinzip Verantwortung: Versuch einer Ethik für die technologische Zivilisation* (4th ed.). Insel.

Kaeser, E. (2019, February 8). Magie und Technik: Warum Handys unser Voodoo sind. *Neue Zuercher Zeitung am Sonntag*. Retrieved March 31, 2025, from https://www.nzz.ch/wissenschaft/wie-unser-glaube-an-magische-kraefte-unsere-beziehung-zu-moderner-technik-praegt-ld.1795923

Kirchschlaeger, P. G. (2017, March 20). Die Rede von 'moral technologies': Eine Kritik aus theologisch-ethischer Sicht. *Feinschwarz*. Retrieved March 31, 2025, from https://www.feinschwarz.net/die-rede-von-moral-technologies/

Kirchschlaeger, P. G. (2021). *Digital Transformation and Ethics. Ethical Considerations on the Robotization and Automation of Society and the Economy and the Use of Artificial Intelligence*. Nomos.

Kosack, S., & Fung, A. (2014). Does transparency improve governance? *Annual Review of Political Science, 17*(1), 65–87. https://doi.org/10.1146/annurev-polisci-032210-144356

Manyika, J., & Bughin, J. (2018, October 15). *The promise and challenge of the age of artificial intelligence*. McKinsey Global Institute. Retrieved March 31, 2025, from https://www.mckinsey.com/featured-insights/artificial-intelligence/the-promise-and-challenge-of-the-age-of-artificial-intelligence

Manyika, J., Chui, M., Miremadi, M., Bughin, J., George, K., Willmott, P., & Dewhurst, M. (2017). *A future that works: Automation, employment, and pro-*

ductivity. McKinsey Global Institute. Retrieved March 31, 2025, from https://www.mckinsey.com/featured-insights/digital-disruption/harnessing-automation-for-a-future-that-works/de-DE

Manzeschke, A., & Karsch, F. (2016). Einleitung. In A. Manzeschke & F. Karsch (Eds.), *Roboter, Computer und Hybride. Was ereignet sich zwischen Menschen und Maschinen? TTN-Studien 5* (pp. 7–14). Nomos.

Marcuse, H. (1964). *One-dimensional man*. Studies in the ideology of advanced industrial society. Beacon Press.

McFee, I. (2024). Feature article: How GenAI will change the world economy. *Economic Outlook, 48*(3), 39–46. https://doi.org/10.1111/1468-0319.12767

McKinsey Global Institute. (2017). *Jobs lost, jobs gained: What the future of work will mean for jobs, skills, and wages*. Retrieved March 31, 2025, from https://www.mckinsey.com/~/media/McKinsey/Industries/Public%20and%20Social%20Sector/Our%20Insights/What%20the%20future%20of%20work%20will%20mean%20for%20jobs%20skills%20and%20wages/MGI-Jobs-Lost-Jobs-Gained-Executive-summary-December-6-2017.pdf

Meyer, M. (2015). Die Digitalisierung als sozialethische Herausforderung. *Kirche und Gesellschaft, 16*(424), 6–7.

Moore, P. V., & Woodcock, J. (2021). Introduction. AI: Making it, faking it, breaking it. In P. V. Moore & J. Woodcock (Eds.), *Augmented exploitation: Artificial intelligence, automation and work* (pp. 1–9). Pluto Press.

Neuhaeuser, C. (2014). Roboter und moralische Verantwortung. In E. Hilgendorf (Ed.), *Robotik im Kontext von Recht und Moral. Robotik und Recht 3* (pp. 269–286). Nomos.

Ohly, L. (2019). *Ethik der Robotik und der Künstlichen Intelligenz* (Theologisch-Philosophische Beiträge zu Gegenwartsfragen (Vol. 22)). Peter Lang.

Panzaru, C., & Negoiță, G. (2022). Artificial intelligence: Challenges and opportunities for the labour market. In C. Larsen, J. Kipper, A. Schmid, & M. Ricceri (Eds.), *The relevance of artificial intelligence in the digital and green transformation of regional and local labour markets across Europe. Perspectives on employment, training, placement, and social inclusion* (pp. 27–44). Nomos.

Paus, E. (2018). The future isn't what it used to be. In E. Paus (Ed.), *Confronting dystopia: The new technological revolution and the future of work* (pp. 1–24). Cornell University Press. https://doi.org/10.7591/9781501719868

Pirvu, R., Mihai, D., Ecobici, N., Antoniu, M., Chirtoc, I. E., & Busan, G. (2024). How does artificial intelligence influence the socio-economic performance of companies in the European Union? *Amfiteatru Economic, 26*(66), 458. https://doi.org/10.24818/ea/2024/66/458

Prettner, K., Geiger, N., & Schwarzer, J. A. (2018). Die Auswirkungen der Automatisierung auf Wachstum. *Beschäftigung und Ungleichheit. Perspektiven Der Wirtschaftspolitik, 19*(2), 59–77. https://doi.org/10.1515/pwp-2018-0017

Restrepo, P. (2023). *Automation: Theory, evidence, and outlook* (Working Paper 31910). National Bureau of Economic Research. https://doi.org/10.3386/w31910

Rueda, D., & Thewissen, S. (2018). Automated but compensated? Technological change and redistribution in advanced democracies. In E. Paus (Ed.), *Confronting dystopia: The new technological revolution and the future of work* (pp. 134–155). Cornell University Press. https://doi.org/10.7591/9781501719868

Scheiber, N. (2025, February 2). Why is this C.E.O. bragging about replacing human with A.I.? *The New York Times*. Retrieved March 31, 2025, from https://www.nytimes.com/2025/02/02/business/klarna-ceo-ai.html

Schweighofer, J. (2016). Zur Befreiung des Menschen von mühevoller Arbeit und Plage durch Maschinen, Roboter und Computer – Auswirkungen der Digitalisierung auf die Arbeitsmärkte. *Wirtschaft und Gesellschaft, 42*(2), 219–255. https://EconPapers.repec.org/RePEc:clr:wugarc:y:2016:42i:2p:219

Schweizerische Nationalkommission Justitia et Pax. (1998). *Machbares Leben? Ethik in der Medizin*. NZN Buchverlag.

Simon, J. (2015). Distributed epistemic responsibility in a hyperconnected era. In L. Floridi (Ed.), *The onlife manifesto. Being human in a hyperconnected era* (pp. 135–151). Springer.

Spielkamp, M. (2019, July 29). Wenn Algorithmen über den Job entscheiden. *Die Presse*. Retrieved March 31, 2025, from https://www.diepresse.com/5662249/wenn-algorithmen-ueber-den-job-entscheiden?from=rss

Wachter, S., Mittelstadt, B., & Russell, C. (2017). Counterfactual explanations without opening the black box: Automated decisions and the GDPR. *Harvard Journal of Law & Technology, 31*(2), 841–887. https://doi.org/10.48550/arXiv.1711.00399

Webb, M. (2019). *The impact of artificial intelligence on the labor market* (SSRN Working Paper No. 3482150). Social Science Research Network. https://doi.org/10.2139/ssrn.3482150

Wilhelms, G. (2018). Wie die Digitalisierung das Verhältnis des Menschen zu seiner Arbeit verändert: Versuch einer Kritik. *AMOS. International, 12*(1), 3. Retrieved March 31, 2025, from https://www.amosinternational.de/magazine/issue-2018-1

Winfield, A. F. T., & Jirotka, M. (2017). The case for an ethical black box. In Y. Gao, S. Fallah, Y. Jin, & C. Lekakou (Eds.), *Towards autonomous robotic systems* (Lecture Notes in Computer Science, Vol. 10454) (pp. 262–273). Springer. https://doi.org/10.1007/978-3-319-64107-2_21

Wong, W. K. O. (2024). The sudden disruptive rise of generative artificial intelligence? An evaluation of their impact on higher education and the global workplace. *Journal of Open Innovation Technology Market and Complexity, 10*(2), 100278. https://doi.org/10.1016/j.joitmc.2024.100278

World Bank. (2016). *World development report 2016: Digital dividends.* World Bank. Retrieved March 31, 2025, from https://www.worldbank.org/en/publication/wdr2016

Wu, Y., & Deng, Y. (2024). Does digital transformation crowd out the employment of lower skill labor? *Bulletin of Economic Research, 76*(3), 726–748. https://doi.org/10.1111/boer.12446

CHAPTER 5

Relief as a Burden: The End of Striving for Full Employment

Abstract In this chapter, the systemic nature of digital transformation of human labor and the impact of data-based systems (DS) on human work is addressed, namely, the end of the strive of the economy for full employment. This ethical analysis takes this systemic changeover seriously by avoiding placing the burden of it on the shoulders of individuals—calling irresponsibly for in-service training and upskilling for a job-market that won't exist anymore—but by developing a systemic response to a systemic transformation. A decoupling of income from work is considered a possible reaction to the massive reduction of paid professional tasks for humans in order to guarantee survival and a life as a human with human dignity for all humans.

Keywords Ethical study • Strive of the economy for full employment • Changeover • In-service training • Upskilling • Job-market • Systemic response • Decoupling of income from work

P. G. Kirchschlaeger, *Ethics and the Digital Transformation of Human Work*, https://doi.org/10.1007/978-3-031-98704-5_5

5.1 Fewer Humans Involved: Fewer Humans Benefit

Sebastian Siemiatkowski, chief executive and a co-founder of the technology-firm Klarna, shakes up "to encourage people, specifically politicians in society, to actually treat this as a serious change that's coming" (Siemiatkowski, as cited in Scheiber, 2025). The reduction of the paid labor-market will provoke a lack of sources of income, of financial means for social welfare, of structuring daily life, of purpose of life, and of the self-understanding of humans. Beyond that, it will contribute to the widening of the gap between rich and poor. All these consequences are of ethical concern due to their impact on justice and on the survival as well as on the human dignity of all humans. "The technological path of AI is not predetermined; with thoughtful action, a future with strong democratic institutions and shared prosperity is possible" (Bell & Korinek, 2023, p. 160). One path towards solutions from an ethical perspective is constituted by addressing the ethical challenges that result from the substantial reduction in paid labor for humans.

As introduced in Chapter 3 "The Uniqueness of Digital Transformation and Data-Based Systems," digital transformation and the use of DS will lead to the core consequence: *Fewer and fewer humans will directly participate and partake in a more efficient and effective value chain* (Kirchschlaeger, 2016a, 2017b). The labeling as "core consequence" from an ethical perspective is justified by the following reasons:

Firstly, it embraces major positive and negative outcomes of digital transformation and DS.

Secondly, it directs the focus on the real challenges of the disruption caused by digital transformation and DS from an ethical perspective. The main challenge is not a lack of financial means because more efficient and more productive value-added chains lead to an increase in that regard, but the question of justice and human rights (Stuebinger, 2019). "A society with cheap robot labor would be an incredibly prosperous one, but we will need to find some way for the vast majority of human beings to share in that prosperity" (Smith, 2013). On the one hand, the distribution of the added value created is at stake or at the center of concern. More is produced but fewer people benefit from it. Even if everything is becoming cheaper due to digital transformation and the use of DS (Worstall, 2012), it does not make any difference if one cannot afford to survive. On the other hand, it is the question of social integration as fewer humans are involved in the value-added chain. "In order to achieve a human-centric scenario for the future, a future in which everyone benefits from automation and the reduction of inherent ethical and inequality issues, just and timely adjustments from all stakeholders and policymakers are needed" (Panzaru & Negoiță, 2022, p. 40). Thirdly, as the core consequence of digital transformation and DS embraces also the replacement of humans from paid professional tasks, fewer working hours and more free time are not necessarily a bad thing. Less work per se does not have to be negative from an ethical standpoint because it allows humans, for example, to devote their time to, e.g., ethically meaningful purposes (among others, the fight against global poverty, the struggle for environmental protection (Burchardt, 2018)). It could also be ethically positive that humans are liberated from the obligation to work (Kellermann, 2014). From an ethical point of view, it is conceivable, among other things, that, depending on its normative orientation, paid work to finance one's livelihood might contribute more to losing, than to finding meaning, might distract humans from the ethically relevant, might create ethically false incentives or coercion, might steal time from other areas of human existence because of work overload (e.g., for idleness, political engagement etc. (Arendt, 2001)), and binds talents and time of humans for activities, which have only one particular purpose that can also be problematic from an ethical point of view. The question of what humans are free to do when they are free from paid professional tasks needs to be discussed. In their role as social actors, religious and worldview-based communities could play an important role in accompanying humans in their quest for the meaning of life, for an image of humans and a self-image, in creating space for this debate, in reflecting upon it from an ethical perspective, in criticizing and participating in the shaping of digital transformation and DS, as well as in the designing of an economy and a society that does not aspire full employment anymore but still a humane and sustainable future for all humans and the planet (Kirchschlaeger, 2016c, 2017a).

5.2 Justice-Oriented and Human Rights-Based Societal and Economic Design

At this point, it seems to be helpful to remind ourselves that the present "work-ethos" was developed approximately 150 years ago and that an adaptation of a "work-ethos" based on a paid job and on full employment is possible, corresponding to the consequences of digital transformation and DS—including a re-valorization of social engagement, social entrepreneurship, "societal time," and spare time. It depends on the design of economic and societal general conditions—the former allowing survival and a life with dignity for all humans. The latter must enable humans to integrate into society and economy without a paid job and allow them to develop a new self-understanding independent from paid professional tasks.

> Ultimately, the impact of the new technological revolution is a political question, not an economic one. A constructive answer to the challenges posed will require forging a social contract in support of a new institutional architecture. That is a long-term and arduous process. Avoiding dystopias will require redistribution on a significant scale, whether of hours worked or the benefits of technological change. Governments, the business sector, unions, and civil society groups have to discuss which aspects of jobs and livelihoods are most challenging given a country's economic and social conditions, and which institutional responses are necessary in the short and the long run. An important step toward starting the dialogue is a greater awareness that the future isn't what it used to be. (Paus, 2018, p. 5)

Ethics should inform this process. From an ethical point of view this implies that, from a macrosocial perspective, the challenge lies not in the amount of financial resources and technological means available, but in the creation of a fair social and economic system (Kirchschlaeger, 2013; Rueda & Thewissen, 2018), *inter alia*, with regard to distributing financial means and the opportunities of participating and partaking, concerning equal opportunities for all, with regard to securing a livelihood for all, and at the same time, upholding entrepreneurial incentives and social peace. Technological progress itself does not lead automatically to more paid professional tasks for humans, higher wages, and more prosperity for all. It depends how the economic and political power over this technological power is allocated (Acemoglu & Johnson, 2023) and how the economic profits are distributed. Beyond that, DS can be used to automate and to monopolize information which damages democracy and democratic participation that already has suffered and is continuing to suffer due to "anti-social media" (Kirchschlaeger, 2024), because humans withdraw from the public space, spend much more time alone (Acemoglu, 2025), and tend to be isolated in their "bubbles" (Gremsl, forthcoming) impeding political deliberations aspiring to master in an ethically sound way the massive reduction of paid professional tasks for humans due to digital transformation and the use of DS.

A justice-oriented and human rights-based societal and economic design of the institutional response to the massive reduction of paid professional tasks becomes even more urgent and imperative amid the fact that the following eight men dominating with their companies digital transformation and the use of DS are part of the ten richest people in the world: Elon Musk (Tesla, SpaceX, and Twitter/X), Jeff Bezos (Amazon), Mark Zuckerberg (Meta/Facebook), Larry Ellison (Oracle), Larry Page, Sergey Brin (Google), and Bill Gates and Steve Ballmer (Microsoft) (Forbes, 2025).

> History shows clearly that when jobs are scarce, the fear of even more unemployment becomes a powerful tool in the hands of politicians and special interests who oppose action on the environment. This has been the case, for example, in those states where coal mining has historically been an important source of jobs, despite the fact that employment in the mining industry has been decimated not by environmental regulation but by mechanization. Corporations with even small numbers of jobs to offer routinely play states and cities against each other, seeking lower taxes, government subsidies, and freedom from regulation. Beyond the United States and other advanced countries, the situation may be far more dangerous. [...] In the worst-case scenario, a combination of widespread economic insecurity, drought, and rising food prices could eventually lead to social and political instability. The greatest risk is that we could face a 'perfect storm' – a situation where technological unemployment and environmental impact unfold in parallel, reinforcing and perhaps even amplifying each other. (Ford, 2015, pp. 284–285)

From an ethical point of view, there is an urgency to act upon it and to create a fair social and economic system.

> The deterioration of the market value of labor in richer parts of the world has been steady over the last forty years. Some poorer states, such as India and China, have enjoyed dramatic increases in the market value of labor during that same period. However, even they are beginning to experience a weakening of labor markets. Globalization has been the driving force of this trend, but the logic of globalization and its more recent ancillary, robotization, will ultimately create a world not of rich and poor nations, but of a unified rich elite and an undifferentiated mass of disenfranchised citizens. For self-protection, that elite will divert the anger and despair of those

> citizens away from policies designed to enhance the market value of labor into the destructive forces of nationalism, xenophobia, and authoritarianism. (Ferraro, 2018, p. 157)

This justice-oriented and human rights-based societal and economic design (Kirchschlaeger, 2013, 2016b) should start with equal opportunities for all so that all humans can be guaranteed survival and a life with human dignity (Kirchschlaeger, 2016d). If the creation of equal opportunities is successful, a "donut-shaped economy" (Keen, 2015) can be avoided, meaning an economic system and a society where the gap between the rich and the poor grows exorbitantly, and the middle of society is eroded.

5.3 The Role of Education

Education is of ethical concern because the aims of education present an ethical question. Primarily, education must take care of not losing sight of the essentials as a result of the pursuit of progress. The same ethical principles, norms, and standards are valid in real and virtual reality. This means also that technological applications should only be introduced or used in education if they make sense from an ethical, from a pedagogical, and from a didactic perspective—not just because they represent innovations or because they are sponsored by, or are a gift from, the private sector. For example, receiving 100 basketballs for free doesn't mean that the school would change its curriculum and fill it with playing basketball, would it? Or at least that is not what should happen even if I love to play Basketball …

At the same time, one must be aware of the fact, that no immediate impact can be expected from education. Adapting education in the face of digital transformation and DS (Jošt et al., 2024; Katsamakas et al., 2024) will only result in medium-term and long-term effects.

The present academic, but also political discourse, about the issue of job-reduction for humans also includes requests to humans to participate in inservice-training preparing themselves for the present and future digital transformation and for the use of DS (Tuffley, 2015), to adapt to the technology-based change (Ellingrud, 2024; Hazan et al., 2024), to make oneself "digitally fit" for the upcoming (Brynjolfsson and McAffee, 2011), and to become ready for a volatile, insecure, complex, and ambiguous reality (Freyth, 2020). For example, the following pattern of argumentation is presented:

> Technological progress is going to leave behind some people, perhaps even a lot of people, as it races ahead. […] there has never been a better time to be a worker with special skills or the right education, because these people can use technology to create and capture value. However, there's never been a worse time to be a worker with only 'ordinary' skills and abilities to offer, because computers, robots, and other digital technologies are acquiring these skills and abilities at an extraordinary rate. (Brynjolfsson and McAffee, 2014 p. 11)

This has been true for technology-based shifts in the past. This is not true anymore for the digital transformation and for the use of DS as explained above because of massive reduction of paid professional tasks for humans, mostly because of the aim of this technological progress—the replacement of humans with machines serving cost-efficiency—and due to the fact that all professional groups are touched by digital transformation and the use of DS.

If a massive reduction of paid professional tasks occurs and is on the horizon of the future, and if digital transformation and DS put an end to an economy striving for full employment, the following reactions are not only as meaningful as writing in the water, but also ethically irresponsible because they distract from the core issues and real challenges and they divert the responsibility for the negative consequences of the massive reduction of paid professional tasks for humans from the powerful to the weaker, namely, from multinational technology-companies to every human seeking employment without hope because the latter, though, does not exist anymore:

From an ethical perspective, it is irresponsible to try to make a fundamental structural problem of the economic system (Di Viggiano, 2018) (massive reduction in paid professional tasks (Ahmed et al., 2020; Burchardt, 2018; Parr, 2022)), an individual challenge (individual responsibility of every human to educate oneself according to the technology-based shift (Adegbite & Govender, 2022; Weber, 2023)).

From an ethical perspective, it is irresponsible to avoid tackling the fundamental change of the economic system consisting in the massive reduction in paid professional tasks for humans with structural measures by calling for reskilling and upskilling as well as for inservice-training in general on an individual level (Cevasco et al., 2019; World Economic Forum [WEF], 2025).

From an ethical perspective, it is irresponsible to request humans to educate themselves when one knows that humans will be replaced anyway. For example, Christoph Franz, Chairman of Roche Holding, states: "We want to ensure that our employees acquire the necessary skills. [...] Most of us will be forced to acquire new skills in our professional lives" (Franz, as cited in Moeckli, 2019, p. 8). Of course, it is positive from an ethical perspective that he claims that his company has a responsibility for this inservice-training of its employees. At the same time, it is ethically problematic to pretend that one can continue with business as usual and to deny the more fundamental challenge that humans—even though they invest in their own education and training—will lose their paid professional tasks anyway.

From an ethical perspective, it is irresponsible for a company to externalize the costs for replacement of humans with machines to the public while requesting humans to educate themselves. For example, the regulatory process by the European Union goes in that direction. Mady Delvaux, at that time a member of the European Parliament, criticized this issue as follows: Lawmakers "refused to take account of possible negative consequences on the job market. They rejected an open-minded and forward-looking debate and thus disregarded the concerns of our citizens" (European Parliament, 2017; Prodhan, 2017).

From an ethical perspective, it is irresponsible to request humans to educate themselves when no one is ready to cover the costs for this inservice-training.

> There is also the financial cost involved and the likelihood of 'pig Cycle' effects: a shortage of pigs (marketable skills) leads to price rises for pigs (wage opportunities for those with the skills), which leads farmers to invest in more pigs, which leads to a glut of pigs and falling prices (wages and job opportunities). Unlike pigs, the skills may have become obsolete by the time they have been expensively obtained. (Standing, 2018, p. 128)

From an ethical perspective, it is irresponsible to request humans to educate themselves when the only two certainties about the future are the central role of technologies and the uncertainty about the future: Nobody knows—due to the elements of the unknown and uncertainty—if any or which inservice-training would be adequate (Celentano, 2021).

From an ethical perspective, it is irresponsible to request humans to train themselves because the pace of technological progress and correspondingly of the disruptive changes of paid professional tasks (if even still existing) is such that humans would run behind continuously reskilling and retraining programs like in a hamster wheel deprived of the possibility of a legitimately autonomous choice and planning of their own professional path and their own life (Belic, 2024).

From an ethical perspective, it is irresponsible when, for example, the role of human resource management in DS-driven market environment is discussed by enumerating the different tasks of human resource management but leaving out the task of accompanying humans faced with the loss of paid professional tasks (Fenwick et al., 2024).

Of course, education plays from an ethical perspective an essential role as a path towards solutions allowing all humans to shape digital transformation and DS and to benefit from technology-based changes and avoiding the ethical risks. Education should, though, neither focus on the fact that the massive reduction of paid professional tasks for humans due to DS will be different regarding tempo (McElheran et al., 2024), timing (Meier, 2024), intensity (Barbieri et al., 2019; Fabian & Hanno, 2019) for different professional tasks, for different skills and competencies as well as for different contexts (as an exemplary study: Tavares et al., 2023), countries (Acemoglu et al., 2020; Apella & Zunino, 2017; Egana del Sol, 2020; Gill et al., 2024; Lladós, 2019), and world regions (Arntz et al., 2016; African Union Commission/Organisation for Economic Co-operation and Development [AUC/OECD] 2021; Grigera and Nava 2021) because, at the end of the day, it will occur everywhere and for everyone (Kirchschlaeger, 2021).

> Rapid and accelerating digitization is likely to bring economic rather than environmental disruption, stemming from the fact that as computers get more powerful, companies have less need for some kinds of workers. (Brynjolfsson & McAfee, 2014, p. 11)

Nor should education lean on the phenomenon that the massive reduction of paid professional tasks for humans due to DS will impact women even more than men, although DS themselves, from a technological standpoint, could avoid gender discrimination in the area of human labor (Oezdemir, 2021). "Across sectors and occupations, underrepresentation of women in managerial positions places them at high risk of displacement by technology" (Brussevich et al. 2019, p. 24).

Education should look the prospect of massive reduction of paid professional tasks for humans in the eye.

From an ethical perspective, the questions need to be addressed more firmly which professional tasks only humans should—normatively speaking, even if humans cannot do it anymore due to digital transformation of human work and the impact of the use of DS on human labor leading to massive reduction of paid professional tasks—contribute exclusively to the economy (Hacker, 2018, pp. 31–60), and in which professional tasks humans outperform by far (if it were not a substantial surpassing, it might be the case that a replacement of humans by machines still occurs due to efficiency gains in other areas). For example, and simply put, machines can only give answers, but they can neither pose questions nor explain to one how they came up with their response.

5.4 Education: Empowering Humans as Humans for Unknown Future

The voices calling for digital skills (Boudjema et al., 2022; Bouwmans et al., 2024; Jaeger & Tewes, 2023; Renz, 2023; Terstegen et al., 2023; Umbach et al., 2020), for DS learning assistants (Blanc et al., 2023), and for the use of DS in education (Hu, 2024) are loud—extremely loud, especially amid the role of digital skills in an environment of massive reduction of paid professional tasks (including those professional tasks requiring digital skills), and amid the fact of lacking scientific studies in favor of the latter two (Linn, 2004, 2024).

As shown in Sect. 1.2 "Data-Based Systems (DS) Rather Than 'Artificial Intelligence'", there are domains of human intelligence where DS already surpass humans today, and this domination will even grow further in the future. At the same time, though, DS are neither able today nor will be able in the future to reach other domains of human intelligence, e.g., emotional and social intelligence as well as moral capability. Therefore, from an ethical perspective but probably also from a pragmatic standpoint, education should be changed to focus on building and fostering capacities, competencies, and skills in these unique characteristics of humans rather than in areas of human intelligence where humans are already today left behind by DS. "If we are going to be able to continue a work-based economy, we need to know that most people can develop better skills that those provided by computers" (Elliott, 2018, p. 44). Education should be changed to empower humans as humans. In other words, this means that education is to transcend just training "technical rationality" but rather "empowering" reason.

This ethically informed change of education should embrace both the structural and systemic dimension as well as the individual dimension of the concrete actors (for example, teachers, pupils, students) (Peters et al., 2018).

Accordingly, education must address questions like the following ones and answer them not from an economic or technophile point of view but based on purely pedagogical and didactic reasoning—in the "best interest of the pupils and students": Should education be digitalized? If so, how far should digitalization of education go? Should every pupil have his/her own tablet or his/her technical device? Or should school be a screen-free oasis? Should school be a place where interpersonal social interaction can be experienced and learned without distraction by screens? Should didactics be digitalized? Should mobile phones be prohibited at school? How should we handle the challenge of having teachers of the twentieth century teaching pupils and students of the twenty-first century? Should education aim neither for analog nor for digital but rather for using exactly what serves its pedagogical, didactic, and ethical purpose and leaving the rest out? Should everyone or anyone be equipped from the point of view of justice and equality? Should teachers be empowered in their self-confidence to be teachers for "millennials" but not to be "millennials" themselves? Should education change the way it deals with the fact that education is changing faster and in shorter intervals? "The belief that education can be replaced by a computer program is a myth. Human contact and mentoring make a significant difference in learning outcomes" (Bossard, 2018), states even Sebastian Thrun, chairman and co-founder of Udacity, a for-profit educational organization offering massive open online courses.

Education should change how it deals with teachers and/or teachers should change with respect to how technology and technological innovation are perceived. Teachers should not be obliged to use technological applications just because they exist, because they are free of charge, because a company links it to a donation for the school, etc. They are obliged to make ethically, pedagogically, and didactically informed and based, rational decisions about whether to use technological applications or not. The criteria of comprehensibility, manageability, and meaningfulness need to be respected. For example, a digital learning assistance system needs to be assessed from an ethical perspective as well (among others, for an assessment of emotions-sensitive training-systems for humans with autism, see Dziobek et al., 2017). This could result in the identification of ethical risks such as dependence, of turning the pupils who are using the technology themselves into a product for providing data, of violations of the human right to privacy and data protection, and so on. For example, Google, Apple, Microsoft, Amazon, and Meta offer modern infrastructure, cloud-services, messenger-services, user-friendly design, and powerful functions to schools under economic pressure—almost for free. But "Google Suite," "School Manager," "Microsoft Office 365," and "Windows 10" represent "spies in the classroom" (Fichter, 2019). They collect as much data as possible from and about children and abuse them economically (Knuth and Ehrenhauser, 2019), while children and their parents erroneously trust the schools and falsely assume that the public education systems are protecting them.

Video conference tools such as "Zoom" collect as much personal data as possible from the hosts and participants of virtual meetings and sell this data to Meta and other third parties (Laaff, 2020).

Beyond that, education should adapt to the digital transformation and DS and prepare adequately for the technology-based change, but even more it should fulfill its design responsibility of digital transformation and DS. Schools are a place where digital transformation and DS must be designed, influenced, guided ethically, changed, and stopped. Education should involve finding ways to enjoy the ethical opportunities of digital transformation and DS. Education should prepare against the ethical risks of digital transformation and DS as well (Schaudel, 2012). For example, the present violations of the human rights to privacy and to data protection should be encountered by educational efforts resulting in "privacy literacy":

> In our opinion, however, privacy does not have to come to an end. However, it is not a matter of course either, but requires first awareness and then action, that control such things. Especially autonomy as a relevant aspect here must also be maintained and mediated. A digital privacy competence seems to be the central means (privacy literacy), such a one but just not with the technical application of media, but rather with an awareness of their semiotic dimension and the constructive character of what is required in

> terms of content and communicated, as only then the application in practice is put on a solid foundation and behavior could be related and reflected upon. In sum, the following skills stand for a privacy competence: a) the knowledge of how media communicate and construct privacy semiotically (medial competence), (b) the ability to reflect on why private data should be considered sensitive (ethical competence), (c) the knowledge of who is collecting, processing, and disclosing private data and for what purpose (structural competence), (d) the assessment of the consequences that from publication private data could result (risk competence), (e) the knowledge of possible (self-)protection measures and privacy protective communication media (action competence), and (f) the empowerment to reflect on power aspects of digitalization – short big data, big power, and big money (systemic analysis and political knowledge). (Grimm & Krah, 2014, pp. 21–22)

Moreover, education must take into account the massive reduction of paid professional tasks for humans in the course of digital transformation and of the use of DS.

Due to digital transformation and DS, it is hard to foresee the economic developments (Brundage, 2015). Education should prepare for the digital transformation and DS strengthening the position of humans in the world, empowering humans to design and steer digital transformation and DS, as well as to deal adequately with uncertainty (Clauson, 2020) and to adapt to the rapid change (Lemke and Monett Diaz, 2020) but also enhancing the unique competencies of humans (García Peñalvo et al., 2024; Lahera Sánchez, 2021; Lanzl et al., 2024; Leyh & Schaeffer, 2024; Margaryan, 2023; Peetz, 2019; Walter, 2024; Webster & Ivanov, 2020), thus distinguishing humans from DS (Brynjolfsson & McAfee, 2014, pp. 10–11; Hasenbein, 2023, pp. 35–55)—e.g., an attitude grounded in human dignity (Pollmann, 2022), human rights (Pollmann, 2022), rights of the child (Bellamy et al., 2007), and sustainability (United Nations [UN] 1987), ethical reason (*Vernunft*) instead of just technical rationality (*Verstand*), critical thinking, the ability to think in complex systems and/or to interact with complex systems, flexibility, social and interpersonal competence (Ellingrud, 2024), creativity (Saint-Germes, 2021; WEF 2025), arts, moral capability, autonomy, ethical competencies, the capability of humans to search, discuss, reflect upon, and define their own purposes of life, the capability of humans to devote working-time, spare time, and "societal time" in accordance with their own horizon of meaning. Education should provide a basis for the ability to take a self-confident standpoint in contexts of uncertainty.

Finally, it is the ethical responsibility of education policy, school principals, teachers, pupils, and students to make schools:

- A place where humans learn about being subjects of responsibility for digital transformation and DS;
- A place where digital transformation and DS are not something given by fate, but objects of human decisions informed also by ethics;
- A place where digital transformation and DS are shaped informed also by ethics;
- A place of critical thinking about everything—including digital transformation and DS and their place in human and in society's lives;
- A place to be strengthened in saying "yes," "yes, but," and "no" to technology-based options from an ethical perspective;
- A place of empowerment for humans so they can live up to and continuously shape rapidly and constantly changing realities and contexts;
- A place to learn to master uncertainty;
- A place with a clear focus on the unique characteristics, competencies, and skills of humans;
- A place contributing to the present and future respect, protection, implementation, and realization of human dignity, human rights, rights of the child, and sustainability.

5.5 Urgency of Action

The societal and economic system is built for an economy striving for full employment. As one can already notice some disruptive steps away from ***doing business with human labor*** towards ***doing business without human labor***, it's urgently time to begin to change our societal and economic system accordingly in order to guarantee a human rights-based and sustainable future for all humans. The urgency stems from the fact that, from a societal and economic point of view, the financial fundament of the existence of individual humans, the tax system, and the instruments of the welfare state like the pension system but also the unemployment insurance, the disability insurance, etc. are built on paid professional tasks for humans. None of these pillars of the societal and economic system are constructed for the situation of a non-working majority of the population which would be the case, though, because of massively reduced paid professional tasks for humans. For example, the unemployment insurance can handle a small non-working minority at maximum but not more than that.

If the societal and economic system is not changed in time, an ethically problematic transitional phase including inhumane living situation, exploding inequality, massively growing gap between rich and poor, social unrest, and political polarization and radicalization will take place because neither the economy, the society, nor the welfare state instruments would be ready for the effects of massive reduction of paid professional tasks for humans due to digital transformation of human work and the impact of the use of DS on human labor.

5.6 Decoupling of Income from Work

Technology-based economic processes lift a burden from humans—namely, the obligation to work—empowering them to fulfill other tasks or to spend their lifetime differently. Fewer working hours can be good news from an ethical point of view, if a society and an economic system are structured accordingly and guarantee all humans a physical survival and a life with human dignity (a life as a human), avoid exclusion as well as inequality (Kristal & Cohen, 2017), enable social ties, and if humans are able to come to a new self-conception, which is mostly independent from paid work.

> Studies show that displaced workers experience sharp drops in earnings and consumption immediately after being displaced, with earnings deficits persisting for years afterward […]. Estimates of wage declines even after re-employment range from 5%–30% for as long as 20 years after displacement […]. Involuntary job loss can also significantly affect physical health, with evidence suggesting that displacement increases mortality risk by 50–100% within the year after separation and by 10–15% annually for the next 20 years. Studies also link job loss to higher rates of depression […], suicide […], alcohol-related disease […], and negative impacts on children's educational attainment […]. Given general-purpose AI's potential to cause job displacement, these findings underscore the importance of policy measures to support affected workers. (AI Action Summit, 2025, p. 114)

Bill Gates, Co-Founder of Microsoft, is in favor of a decoupling of income and work (Juergens & Pletter, 2025, pp. 19–20) for the following reasons:

> In the abstract, artificial intelligence will make us more productive and reduce the cost of office work because the software will work autonomously. and when the software is connected to robots, it will also affect factory workers. [...] We can produce more or take more time off. That sounds good at first. But if it happens quickly, this change will be very disruptive. What does this mean for the tax system, for example? We have organized our society around work. (Juergens & Pletter, 2025, p. 20, translated by the author)

While maintaining entrepreneurial incentives and thus preventing demotivation, a decoupling of income from work could therefore be considered in order to guarantee survival and a life as a human with human dignity for all humans and to enable social peace.

References

Acemoglu, D. (2025, February 1). Die Demokratie steckt weltweit in der Krise. *Schweiz am Wochenende Limmattal*, pp. 12–13.

Acemoglu, D., & Johnson, S. (2023). *Power and progress: Our thousand-year struggle over technology and prosperity*. PublicAffair.

Acemoglu, D., LeLarge, C., & Restrepo, P. (2020). Competing with robots: Firm-level evidence from France. *AEA Papers and Proceedings, 110*, 383–388. https://doi.org/10.1257/pandp.20201003

Adegbite, W., & Govender, C. (2022). How Fourth Industrial Revolution skillsets mediate the relationship between work integrated learning, graduate employability, and future job. *EUREKA Social and Humanities, 4*, 3–17. https://doi.org/10.21303/2504-5571.2022.002425

Ahmed, A. A. A., Saha, S., & Donepudi, P. K. (2020). Emerging market economy and artificial intelligence: Consequences for the future of jobs. *PalArch's Journal of Archaeology of Egypt/Egyptology, 17*(6), 5562–5574. https://doi.org/10.5281/zenodo.5562662

AI Action Summit. (2025). *International AI safety report. The international scientific report on the safety of advanced AI*. Retrieved March 31, 2025, from https://www.gov.uk/government/publications/international-ai-safety-report-2025

Apella, I., & Zunino, G. (2017, October 12). *Technological change and the labor market in Argentina and Uruguay: A task content analysis*. World Bank Policy Research Working Paper No. 8215. National Bureau of Economic Research. Retrieved March 31, 2025 from https://ssrn.com/abstract=3051995

Arendt, H. (2001). *Vita activa oder vom tätigen Leben* (Original work published 1960). Piper.

Arntz, M., Gregory, T., & Zierahn, U. (2016). *The risk of automation for jobs in OECD countries: A comparative analysis* (OECD Social, Employment and Migration Working Paper 189). OECD Publishing. https://doi.org/10.1787/5jlz9h56dvq7-en

AUC/OECD. (2021). *Africa's development dynamics 2021: Digital transformation for quality jobs.* OECD Publishing. https://doi.org/10.1787/0a5c9314-en

Barbieri, L., Mussida, C., Piva, M., & Vivarelli, M. (2019). *Testing the employment impact of automation, robots and AI: A survey and some methodological issues* (IZA Discussion Paper No. 12612). IZA Institute of Labor Economics. https://doi.org/10.2139/ssrn.3457656

Belic, J. (2024). Institutions, automation, and legitimate expectations. *The Journal of Ethics, 28*(3), 505–525. https://doi.org/10.1007/s10892-023-09440-x

Bell, S. A., & Korinek, A. (2023). AI's economic peril. *Journal of Democracy, 34*(4), 151–161. https://doi.org/10.1353/jod.2023.a907696

Bellamy, C., Zermatten, J., Kirchschlaeger, P. G., & Kirchschlaeger, T. (Eds.). (2007). *Realizing the rights of the child* (Swiss Human Rights Book 2). Rueffer & Rub.

Blanc, B., Feichtenbeiner, R., Beudt, S., & Pinkwart, N. (2023). AI in vocational rehabilitation – Intelligent assistance for people with disabilities. In I. Knappertusch & K. Gondlach (Eds.), *Work and AI 2030* (pp. 373–381). Springer. https://doi.org/10.1007/978-3-658-40232-7_41

Bossard, C. (2018, March 9). Digital first! Reflexion second. *Luzerner Zeitung.* Retrieved March 31, 2025, from https://www.luzernerzeitung.ch/schweiz/digital-first-reflexion-second-ld.81789

Boudjema, K., Simon, P., Moulin, T., Pon, D., Chays, A., & Vouhé, P. (2022). Bonnes pratiques en matière de télémédecine. *Bulletin De L'Académie Nationale De Médecine, 206*(5), 657–659. https://doi.org/10.1016/j.banm.2022.03.007

Bouwmans, M., Lub, X., Orlowski, M., & Nguyen, T. (2024). Developing the digital transformation skills framework: A systematic literature review approach. *PLoS ONE, 19*(7), e0304127. https://doi.org/10.1371/journal.pone.0304127

Brundage, M. (2015, January 25). Economic possibilities for our children: Artificial intelligence and the future of work, education, and leisure. *AAAI Workshop: Artificial Intelligence and Ethics 2015*, Austin, Texas, USA. Retrieved March 31, 2025, from https://dblp.org/db/conf/aaai/ethics2015.html

Brussevich, M., Dabla-Norris, M. E., & Khalid, S. (2019). *Is technology widening the gender gap? Automation and the future of female employment.* International Monetary Fund.

Brynjolfsson, E., & McAfee, A. (2011). *Race against the machine: How the digital revolution is accelerating innovation, driving productivity, and irreversibly transforming employment and the economy.* Digital Frontier Press.

Brynjolfsson, E., & McAfee, A. (2014). *The second machine age: Work, progress, and prosperity in a time of brilliant technologies*. W. W. Norton & Company.

Burchardt, A. (2018). So schnell schafft der Mensch sich nicht ab! In Konrad-Adenauer-Stiftung (Ed.), *Zwischenruf. Die digitale Zukunft*. Retrieved March 31, 2025, from https://www.kas.de/de/web/auslandsinformationen/artikel/detail/-/content/so-schnell-schafft-der-mensch-sich-nicht-ab-

Celentano, D. (2021). Work justice beyond redistribution and recognition. In D. Celentano & L. Caranti (Eds.), *Paradigms of justice* (pp. 90–117). Routledge.

Cevasco, L. J., Corvalán, J. G., & Le Fevre Cervini, E. M. (2019). *Intelligenza artificiale e lavoro. Costruire un nuovo paradigma occupazionale*. Edizioni di Comunità.

Clauson, M. (2020). The future of work. In B. J. Hoffman, M. K. Shoss, & L. A. Wegman (Eds.), *The Cambridge handbook of the changing nature of work* (pp. 555–582). Cambridge University Press.

Di Viggiano, P. L. (2018). Etica, robotica e lavoro: Profili d'informatica giuridica. *Revista Opinião Jurídica, 16*(22), 247–266. https://doi.org/10.12662/2447-6641oj.v16i22.p247-266.2018

Dziobek, I., Lucke, U., & Manzeschke, A. (2017). Emotions-sensitive Trainingssysteme für Menschen mit Autismus. In M. Eibl & M. Gaedke (Eds.), *INFORMATIK 2017* (pp. 369–380). Gesellschaft für Informatik.

Egana del Sol, P. (2020). *The future of work in developing economies: What can we learn from the south?* GLO Discussion Paper, No. 483. Global Labor Organization (GLO). Retrieved March 31, 2025, from https://www.econstor.eu/handle/10419/214652

Ellingrud, K. (2024, January 30). *How the US job landscape is changing – And how to adapt*. McKinsey Global Institute. Retrieved March 31, 2025, from https://www.mckinsey.com/mgi/overview/in-the-news/how-the-us-job-landscape-is-changing-and-how-to-adapt

Elliott, S. W. (2018). Artificial intelligence, robots, and work: Is this time different? *Issues in Science and Technology, 35*(1), 40–44. Retrieved March 31, 2025, from https://issues.org/artificial-intelligence-robots-and-work-is-this-time-different/

European Parliament. (2017, January 12). *Rise of the robots: Mady Delvaux on why their use should be regulated*. European Parliament. Retrieved March 31, 2025, from https://www.eumonitor.eu/9353000/1/j9vvik7m1c3gyxp/vkatp3auh9x3?ctx=vg9pjpw5wsz1&v=1&tab=1&start_tab0=590

Fabian, S., & Hanno L. (2019). *Back to the future – Changing job profiles in the digital age*. EconStor Preprints. ZBW – Leibniz Information Centre for Economics. Preprint Retrieved March 31, 2025, from https://www.econstor.eu/handle/10419/202035

Fenwick, A., Molnar, G., & Frangos, P. (2024). The critical role of HRM in AI-driven digital transformation: A paradigm shift to enable firms to move from AI implementation to human-centric adoption. *Discover Artificial Intelligence, 4*(1), 34–16. https://doi.org/10.1007/s44163-024-00125-4

Ferraro, V. (2018). The crisis of the liberal international order: Technological change and the rise of the right. In E. Paus (Ed.), *Confronting dystopia: The new technological revolution and the future of work* (pp. 156–178). Cornell University Press. https://doi.org/10.7591/9781501719868

Fichter, A. (2019, July 2). Der Spion im Schulzimmer. *Republik*. https://www.republik.ch/2019/07/02/der-spion-im-schulzimmer

Forbes. (2025). *The real-time billionaires list*. Forbes. Retrieved March 31, 2025, from https://www.forbes.com/real-time-billionaires/#56bc7c7f3d78

Ford, M. (2015). *Rise of the robots: Technology and the threat of a jobless future*. Basic Books.

Freyth, A. (2020). *Veränderungsbereitschaft stärken. Impulse und Übungen für Mitarbeiter und Führungskräfte*. Springer.

García Peñalvo, F. J., Llorens-Largo, F., & Vidal, J. (2024). La nueva realidad de la educación ante los avances de la inteligencia artificial generativa. *RIED-Revista Iberoamericana de Educación a Distancia, 27*(1), 9–39. https://doi.org/10.5944/ried.27.1.37716

Gill, A., Heggart, K., Johns, A., Knight, S., Narayan, B., & Shibani, A. (2024). *Submission in response to the parliamentary inquiry into digital transformation of workplaces* (Submission No. 12). House Standing Committee on Employment, Education and Training. Retrieved March 31, 2025, from https://www.unsw.edu.au/engineering/research-technology/research-centres-institutes/trustworthy-digital-society

Gremsl, T. (forthcoming). *Zwischen digitaler Spaltung und neuer Agora. Eine sozialethische Analyse des Einflusses sozialer Medien in liberal-pluralen Demokratien*.

Grigera, J., & Nava, A. (2021). El futuro del trabajo en América Latina: crisis, cambio tecnológico y control. *El Trimestre Económico, 88*(352), 1011–1042. https://doi.org/10.20430/ete.v88i352.1242

Grimm, P., & Krah, H. (2014, October 29). *Ende der Privatheit? Eine Sicht der Medien- und Kommunikationswissenschaft*. Institut für Digitale Ethik. Retrieved March 31, 2025, from https://www.zbw-mediatalk.eu/wp-content/uploads/2016/08/Ende_der_Privatheit_Grimm_Krah.pdf

Hacker, W. (2018). *Menschengerechtes Arbeiten in der digitalisierten Welt: Eine wissenschaftliche Handreichung* (Buchreihe Mensch – Technik -Organisation / Band 49). vdf.

Hasenbein, M. (2023). Wandel der Arbeits- und Organisationswelt durch Künstliche Intelligenz. In M. Hasenbein (Ed.), *Mensch und KI in Organisationen* (pp. 35–55). Springer. https://doi.org/10.1007/978-3-662-66375-2_4

Hazan, E., Madgavkar, A., Chui, M., Smit, S., Maor, D., Dandona, G. S., & Huyghues-Despointes, R. (2024, May 21). *A new future of work: The race to deploy AI and raise skills in Europe and beyond*. McKinsey Global Institute. Retrieved March 31, 2025, from https://www.mckinsey.com/mgi/our-research/a-new-future-of-work-the-race-to-deploy-ai-and-raise-skills-in-europe-and-beyond

Hu, Y.-H. (2024). Implementing generative AI chatbots as a decision aid for enhanced values clarification exercises in online business ethics education. *Educational Technology & Society, 27*(3), 356–373. https://www.jstor.org/stable/48787035

Jaeger, C., & Tewes, S. (2023). AI in continuing education of the future. In I. Knappertusch & K. Gondlach (Eds.), *Work and AI 2030* (pp. 361–371). Springer. https://doi.org/10.1007/978-3-658-40232-7_40

Jošt, G., Taneski, V., & Karakatič, S. (2024). The impact of large language models on programming education and student learning outcomes. *Applied Sciences, 14*(10), 4115. https://doi.org/10.3390/app14104115

Juergens, J., & Pletter, R. (2025, February 5). Trump gab mir seine Nummer: Ich werde anrufen, wenn ich verhindern kann, dass Millionen sterben. *Die Zeit*. https://www.zeit.de/2025/06/bill-gates-microsoft-usa-technologie-entwicklung

Katsamakas, E., Pavlov, O. V., & Saklad, R. (2024). Artificial intelligence and the transformation of higher education institutions: A systems approach. *Sustainability, 16*(14), 6118. https://doi.org/10.3390/su16146118

Keen, A. (2015). *Das digitale Debakel*. Deutsche Verlags-Anstalt (DVA).

Kellermann, P. (2014, November 13). Werden Roboter uns das Grundeinkommen bringen? *Der Standard*. https://www.derstandard.at/story/2000008033740/werden-roboter-uns-das-grundeinkommen-bringen

Kirchschlaeger, P. G. (2013). Gerechtigkeit und ihre christlich-sozialethische Relevanz. *Zeitschrift für katholische Theologie, 135*(4), 433–456. https://www.jstor.org/stable/i24160833

Kirchschlaeger, P. G. (2016a, March 30). Digitalisierung und Robotisierung der Gesellschaft aus theologisch-ethischer Perspektive. *Feinschwarz.net*. Retrieved March 31, 2025, from https://www.feinschwarz.net/digitalisierung-und-robotisierung-der-gesellschaft-aus-theologisch-ethischer-perspektive/

Kirchschlaeger, P. G. (2016b). Globale Gerechtigkeit aus einer finanzethischen Perspektive. *Freiburger Zeitschrift für Philosophie und Theologie, 63*(2), 534–552.

Kirchschlaeger, P. G. (2016c). *Menschenrechte und Religionen. Nichtstaatliche Akteure und ihr Verhältnis zu den Menschenrechten* (Gesellschaft – Ethik – Religion (Vol. 7)). Ferdinand Schoeningh.

Kirchschlaeger, P. G. (2016d). Missachtung der Menschenwürde als Schlüsselerfahrung. In K. Kraemer & K. Vellguth (Eds.), *Menschenwürde. Diskurse zur Universalität und Unveräusserlichkeit* (Theologie der Einen Welt 8) (pp. 193–206). Herder.

Kirchschlaeger, P. G. (2017a). Religionen als moralische Akteure. In R. Bergold, J. Sautermeister, & A. Schroeder (Eds.), *Dem Wandel eine menschliche Gestalt geben. Sozialethische Perspektiven für die Gesellschaft von morgen. Festschrift zur Neueröffnung und zum 70-jährigen Bestehen des Katholisch-Sozialen Instituts* (pp. 133–158). Herder.

Kirchschlaeger, P. G. (2017b). Roboter und Ethik. *Aktuelle Juristische Praxis, 26*(2), 240–249.

Kirchschlaeger, P. G. (2021). *Digital Transformation and Ethics. Ethical Considerations on the Robotization and Automation of Society and the Economy and the Use of Artificial Intelligence*. Nomos.

Kirchschlaeger, P. G. (2024, December 2). Protecting children from anti-social media. *Project Syndicate*. Retrieved March 31, 2025, from https://www.project-syndicate.org/commentary/australia-ban-on-children-using-social-media-should-be-emulated-by-peter-g-kirchschlager-2024-12

Knuth, H., & Ehrenhauser, A. (2019, August 12). Wer darf mitmachen? *Die Zeit* (33), 55–56. Retrieved March 31, 2025, from https://www.zeit.de/2019/33/digitalisierung-schule-it-konzerne-kooperation

Kristal, T., & Cohen, Y. (2017). The causes of rising wage inequality: The race between institutions and technology. *Socio-Economic Review, 15*(1), 187–212. https://doi.org/10.1093/ser/mww006

Laaff, M. (2020, March 31). Ok, Zoomer. *Die Zeit*. Retrieved March 31, 2025, from https://www.zeit.de/digital/2020-03/videokonferenzen-zoom-app-homeoffice-quarantaene-coronavirus/komplettansicht

Lahera Sánchez, A. (2021). El debate sobre la digitalización y la robotización del trabajo (humano) del futuro: automatización de sustitución, pragmatismo tecnológico, automatización de integración y heteromatización. *Revista Española De Sociología, 30*(3), a66. https://doi.org/10.22325/fes/res.2021.66

Lanzl, J., Schnaak, F., Schoettl, F., & Gimpel, H. (2024). Der Einfluss von digitalen Technologien auf Wissensarbeit: Kompetenzen im Wandel. *HMD Praxis Der Wirtschaftsinformatik, 61*(1), 141–158. https://doi.org/10.1365/s40702-023-01031-6

Lemke, C., & Monett Diaz, D. (2020). KI und Automatisierung: Wirkung auf die Arbeit der Zukunft. In J. Nachtwei & A. Sureth (Eds.), *Sonderband Zukunft der Arbeit, Human Resources Consulting Review* (Vol. 12, pp. 400–403). VQP. https://doi.org/10.13140/RG.2.2.33923.32803

Leyh, C., & Schaeffer, T. (2024). Digitale kompetenzen. *HMD Praxis Der Wirtschaftsinformatik, 61*(1), 1–5. https://doi.org/10.1365/s40702-024-01043-w

Linn, S. (2004). *Consuming kids. The hostile takeover of childhood*. The New Press.

Linn, S. (2024, March 13). *Consuming kids! Digital transformation and the right of the child* [Oral conference contribution]. International Conference "Right to a Future, Rights of a Child", Loyola University Chicago & Institute of Social Ethics ISE of the University of Lucerne, Lucerne, Switzerland.

Lladós, J. (2019). Revolució 4.0: progrés o precarització? *Oikonomics. Revista d'economia, empresa i societat, 12*, 1–3. Retrieved March 31, 2025, from https://raco.cat/index.php/oikonomics/issue/view/30344

Margaryan, A. (2023). Artificial intelligence and skills in the workplace: An integrative research agenda. *Big Data & Society, 10*(2). https://doi.org/10.1177/20539517231206804

McElheran, K., Li, J. F., Brynjolfsson, E., Kroff, Z., Dinlersoz, E., Foster, L., & Zolas, N. (2024). AI adoption in America: Who, what, and where. *Journal of Economics & Management Strategy, 33*(2), 375–415. https://doi.org/10.1111/jems.12576

Meier, D. (2024). *The Future of work: Ethical evaluation of the change of human labor in the context of advancing automation*. Nomos.

Moeckli, A. (2019, February 11). Roche-Präsident über Digitalisierung: "Ich halte nichts von Angstmacherei." *Luzerner Zeitung*. Retrieved March 31, 2025, from https://www.luzernerzeitung.ch/wirtschaft/roche-praesident-ueber-digitalisierung-ich-halte-nichts-von-angstmacherei-ld.1092784

Oezdemir, E. D. (2021). General evaluation for the future of women employment in the digital transformation. In M. Ozsahin (Ed.), *New strategic, social and economic challenges in the age of society 5.0: Implications for sustainability* (Vol. 121, pp. 30–35). European Publisher. https://doi.org/10.15405/epsbs.2021.12.04.4

Panzaru, C., & Negoiță, G. (2022). Artificial intelligence: Challenges and opportunities for the labour market. In C. Larsen, J. Kipper, A. Schmid, & M. Ricceri (Eds.), *The relevance of artificial intelligence in the digital and green transformation of regional and local labour markets across Europe. Perspectives on employment, training, placement, and social inclusion* (pp. 27–44). Nomos. https://doi.org/10.5771/9783957104113

Parr, T. (2022). Automation, unemployment, and insurance. *Ethics and Information Technology, 24*(3). https://doi.org/10.1007/s10676-022-09639-w

Paus, E. (2018). The future isn't what it used to be. In E. Paus (Ed.), *Confronting dystopia: The new technological revolution and the future of work* (pp. 1–24). Cornell University Press. https://doi.org/10.7591/9781501719868

Peetz, D. (2019). Digitalisation and the jobs of the future. In D. Peetz (Ed.), *The realities and futures of work* (pp. 83–112). ANU Press. https://doi.org/10.22459/RFW.2019

Peters, M. A., Jandrić, P., & Hayes, S. (2018). The curious promise of educationalising technological unemployment: What can places of learning really do about the future of work? *Educational Philosophy and Theory, 51*(3), 242–254. https://www.tandfonline.com/doi/full/10.1080/00131857.2018.1439376

Pollmann, A. (2022). *Menschenrechte und Menschenwürde. Zur philosophischen Bedeutung eines revolutionären Projekts*. Suhrkamp.

Prodhan, G. (2017, February 17). European parliament calls for robot law, rejects robot tax. *Reuters*. Retrieved March 31, 2025, from https://www.reuters.com/article/us-europe-robots-lawmaking/european-parliament-calls-for-robot-law-rejects-robot-tax-idUSKBN15V2KM

Renz, A. (2023). AI in education: Educational technology and AI. In I. Knappertusch & K. Gondlach (Eds.), *Work and AI 2030* (pp. 353–360). Springer. https://doi.org/10.1007/978-3-658-40232-7_39

Rueda, D., & Thewissen, S. (2018). Automated but compensated? Technological change and redistribution in advanced democracies. In E. Paus (Ed.), *Confronting dystopia: The new technological revolution and the future of work* (pp. 134–155). Cornell University Press. https://doi.org/10.7591/9781501719868

Saint-Germes, È. (2021). La créativité vue comme une compétence. Enjeux et prospective en matière d'employabilité. *Revue d'économie industrielle, 174*, 187–222. https://doi.org/10.4000/rei.10310

Schaudel, D. (2012). Ethik und/oder/wegen Automatisierung?! *Automatisierungstechnik, 60*(1), 4–9. https://doi.org/10.1524/auto.2012.0972

Scheiber, N. (2025, February 2). Why is this C.E.O. bragging about replacing human with A.I.? *The New York Times*. Retrieved March 31, 2025, from https://www.nytimes.com/2025/02/02/business/klarna-ceo-ai.html

Smith, N. (2013, January 14). The end of labor: How to protect workers from the rise of robots. *The Atlantic*. Retrieved March 31, 2025, from https://www.theatlantic.com/business/archive/2013/01/the-end-of-labor-how-to-protect-workers-from-the-rise-of-robots/267135/

Standing, G. (2018). Taskers in the precariat: Confronting an emerging dystopia. In E. Paus (Ed.), *Confronting dystopia: The new technological revolution and the future of work* (pp. 115–133). Cornell University Press.

Stuebinger, E. (2019). Die Digitalisierung der Arbeitswelt als wirtschafts- und sozialethische Herausforderung. *Zeitschrift Für Wirtschafts- Und Unternehmensethik, 20*(2), 202–230. https://doi.org/10.5771/1439-880x-2019-2-202

Tavares, C., Oliveira, L., Duarte, P., & Da Silva, M. M. (2023). Artificial intelligence: A blessing or a threat for language service providers in Portugal. *Informatics, 10*(4), 81. https://doi.org/10.3390/informatics10040081

Terstegen, S., Schmalen, B., Hinz, A., & Pricelius, M. (2023). Introductory qualification on artificial intelligence: Productive and humane work design with AI in small and medium-sized enterprises. In I. Knappertusch & K. Gondlach (Eds.), *Work and AI 2030* (pp. 343–351). Springer. https://doi.org/10.1007/978-3-658-40232-7_38

Tuffley, D. (2015, January 1). Job survival in the age of robots and intelligent machines. *The Conversation*. Retrieved March 31, 2025, from https://theconversation.com/job-survival-in-the-age-of-robots-and-intelligent-machines-33906

Umbach, S., Haberzeth, E., Boeving, H., & Glaß, E. (2020). *Kompetenzver schiebungen im Digitalisierungsprozess: Veränderungen für Arbeit und Weiterbildung aus Sicht der Beschäftigten*. W. Bertelsmann Verlag.

UN. (1987). *Report of the world commission on environment and development: Our common future*. Oxford University Press. Retrieved March 31, 2025, from https://sustainabledevelopment.un.org/content/documents/5987our-common-future.pdf

Walter, Y. (2024). Embracing the future of artificial intelligence in the classroom: The relevance of AI literacy, prompt engineering, and critical thinking in modern education. *International Journal of Educational Technology in Higher Education, 21*(1), 15–29. https://doi.org/10.1186/s41239-024-00448-3

Weber, E. (2023). Digitalisation and regulation in the labour market. *List Forum Für Wirtschafts- Und Finanzpolitik, 50*(1–2), 107–117. https://doi.org/10.1007/s41025-023-00253-2

Webster, C., & Ivanov, S. (2020). Robotics, artificial intelligence, and the evolving nature of work. In B. George & J. Paul (Eds.), *Digital transformation in business and society* (pp. 127–143). Springer. https://doi.org/10.1007/978-3-030-08277-2_8

World Economic Forum. (2025, January 7). *The Future of Jobs Report 2025*. Retrieved March 31, 2025 from https://www.weforum.org/publications/the-future-of-jobs-report-2025/

Worstall, T. (2012, December 10). That robot economy and the rentier class. *Forbes*. Retrieved March 31, 2025 from https://www.forbes.com/sites/timworstall/2012/12/10/that-robot-economy-and-the-rentier-class/#4e678294763f

CHAPTER 6

Which Functions Does a Paid Job Fulfill?

Abstract In this chapter, the functions of paid professional tasks are identified. Paid professional tasks cover aspects which go far beyond, for example, providing financial security and guaranteeing physical survival or financing the social welfare system. Forty functions can be identified which paid professional tasks fulfill, like—among others—structuring daily life, providing purpose of life or identity for humans, serving as a source of self-confidence, and allowing oneself to grow into a society by enabling one to learn to coexist and to arrange oneself with other humans one might not have had the chance to choose freely and, this way, interconnect and become part of a society.

Keywords Functions of paid professional tasks • Financial security • Physical survival • Structuring daily life • Self-confidence • Coexistence • Interconnection • Integration of society

When one starts to approach the option to decouple income from work, the question should be addressed from an ethical perspective about which functions paid professional tasks—that would fall away in such a scenario of decoupling income from work—fulfill for humans.

P. G. Kirchschlaeger, *Ethics and the Digital Transformation of Human Work*, https://doi.org/10.1007/978-3-031-98704-5_6

For humans, paid professional tasks go beyond salary. They possess more meaning than procuring financial security (Gheaus & Herzog, 2016). Analyzing current paid professional tasks for humans and their functions that are of relevance from an ethical standpoint while avoiding a glorification and mystification of paid human labor, while taking a differentiated look at paid professional tasks by always understanding work as an interaction with quiet time and recreation, but by keeping in mind that labor exists for humans and not the other way around, the following functions, among others, can be identified:

1. To open up the possibility to participate in shaping the world: Paid professional tasks allow one to explore and realize one's own possibilities to craft the reality one is living in together with other humans.
2. To create the option to contribute to a higher, ethically sound purpose: They allow one to become part of something with a deeper meaning or something which is ethically right or good.
3. To contribute to peaceful coexistence in solidarity: They allow one to work together with others in a respectful and constructive manner and, this way, to become part of something which fosters and promotes the tolerant and considerate living together in a society.
4. To enable participation in and sharing of the economic added value: They allow one to provide one's own part to and to benefit from the chain of creation of added value.
5. To act as a source of meaning: They allow one to find a purpose for one's own life.
6. To serve as a source of self-confidence: They allow one to gain trust in oneself.
7. To give support: It allows one to help someone else or others.
8. To influence the image of humans: They allow one to contribute to the self-understanding of humans.
9. To create, design, and realize one's own life: They allow one to determine in parts one's own existence.
10. To contribute to the development and unfolding of a self-conception: They allow one to craft one's own understanding of oneself.
11. To create identity: They allow one to draft one of our identities (Kirchschlaeger, 2023). The following example can illustrate the meaning of this function: If one goes to a party and does not know

anyone, the question might be posed "What are you doing in life?", usually the answer is not "I love skiing, snowboarding, and surfing" but rather what one's job is, or which study program or education one is in to reach a qualification for a certain job profile, or—if one is retired—what one's job was before the retirement.

12. To open up self-fulfillment: They allow one to accomplish a task and the path towards it, as well as to enjoy positive emotions and feelings reaching a goal can transmit.
13. To enable self-efficacy: They allow one to explore and realize one's own effectiveness.
14. To further self-determination: They allow one to enjoy more autonomy and freedom to make choices for oneself.
15. To contribute to the structuring of daily life: They allow one to know why to get up in the morning, why to go to bed in the evening, or why to relax and rest on the weekend.
16. To enjoy the recognition of the own performance: They allow one to perceive the results of one's own efforts.
17. To enjoy the monetary recognition of the fulfillment of paid professional tasks: They allow one to perceive that one's own services are valued also economically and financially.
18. To enjoy the accomplishment of a paid professional task: They allow one to fulfill an economically and financially binding task, to live up to this economic and financial duty, and to perceive the positive emotions and feelings the path itself towards an end as well as reaching the goal can transmit.
19. To create the possibility to be self-sufficient: They allow one to experience that one can take care of oneself autarkically.
20. To not be alone and isolated: They allow one to leave one's own home, private sphere, "nest," or "bubble," and to join others in a certain common activity in an economically and financially incentivized setting.
21. To be listened to: They allow one to expose oneself to others and to be heard.
22. To receive praise, recognition, and affirmation: They allow one the possibility to receive feedback for one's own contributions.
23. To experience success and failure and learn how to handle them: They allow one to experience positive or negative results of one's own efforts as well as the corresponding positive or negative feed-

back, and to develop an ability to receive it well and to deal with it in a constructive way.

24. To be trusted with influence, to gain and use scope for decision-making and action: They allow one to experience that someone else puts a task and corresponding choices in one's hands and is willing to pay for it.
25. To develop and educate oneself and to learn: They allow one to improve one's own weaknesses and to achieve a mentality that is open for new insights and findings and to gain hands-on experience, know-how, and knowledge.
26. To experience the promotion of one's own talents and the sharpening of one's own abilities: They allow one to foster one's own strengths and to improve even further one's own main capacities.
27. To develop one's personality: They allow one to grow as a human being in another part of one's own existence apart from one's own private life.
28. To grow into a certain role and fill it: They allow one to live up to the requirements, duties, and responsibilities of a given task in an economically and financially incentivized setting.
29. To help build self-esteem: They allow one to enjoy the support in developing one's own positive picture of oneself by the accomplishment of tasks recognized somehow economically and financially.
30. To grow into a society: They allow one to learn to coexist and to arrange oneself with other humans one might not have had the chance to choose freely and, this way, interconnect and become part of a society. In order to illustrate this point concretely, the pursuit of a paid professional task can come with the challenge that one must work together with other humans one did not choose, and one might not even like or even find unsympathetic which plays an integrative and inclusive role for the cohesion of a society.
31. To further social inclusion: They allow one to join and to become an active part of a society and, by this, to integrate into a society which promotes an open society growing together.
32. To support or cause the creation of interpersonal relationships and networks: They allow one to get to know other humans, organizations, institutions, associations, groups, and networks. The following question can provoke a concrete illustration of this function: If

you imagine all your friends, colleagues, and people you know, which ones did you meet and get to know via paid professional tasks or via education, further education, or inservice-training on the way to paid professional tasks?

33. To contribute to the building or growing of society:

> All those who are gainfully employed and can live from their income feel valued as equal citizens: they can contribute something to society that is considered so valuable that they can support themselves from the compensation they receive in return.
>
> Gainful employment, which is primarily organized in the form of dependent employment, is therefore highly charged normatively, although two aspects of lack of freedom are constitutive for employment relationships. Employees must work to secure their livelihood and that of their family, and they are subject to the employer's right to issue instructions during working hours. (Emunds, 2017, p. 33, translated by the author)

34. To get to know and be challenged by reality: They allow one to step out in the world, to be exposed to contexts and status quo one would have not chosen to be exposed to on a voluntary basis, and to deal with them.
35. To learn or improve mastering of the distinction between important and unimportant things and how to prioritize: They allow one to distinguish the fundamental from the superficial, the urgent from the not pressing, as well as to mature in decision-making and in setting agendas.
36. To enjoy leisure time, holidays, and vacations differentiated from working hours: They allow one to celebrate the unique quality of time without obligations and without sense because of the absence of professional obligation, duty, and responsibility.
37. To enjoy the private life and networks differentiated from the professional life and networks: They allow one to celebrate the unique quality of the private as the key to it lies in its difference from the professional sphere.
38. To potentially create incentives for entrepreneurship: They allow one to strive for self-made success.
39. To potentially trigger incentives for innovation: They allow one to strive for exploring the undiscovered.

40. To contribute to a just, fair, and non-discriminatory society by creating access to paid professional tasks designed in such a way that there are equal opportunities: They allow one—without any difference—to apply for an opportunity to get involved in an economically and financially incentivized setting.

If one were decoupling income from work, what would happen to these functions of paid professional tasks which are of ethical relevance? What would happen to these ethically positive functions of paid professional tasks which demonstrate that paid professional tasks go beyond their economic and financial meaning? If one were decoupling income from work, what would happen to these opportunities of ethically positive purposes provided by paid professional tasks?

One of the economic concepts encompassing the idea of decoupling income from work represents the unconditional basic income-approach or the universal basic income-approach. Therefore, the question can be asked from an ethical perspective about which effect the unconditional basic income-approach or the universal basic income-approach with its separation of income and labor causes on these 40 functions of current paid jobs.

References

Emunds, B. (2017). Weniger Arbeit, mehr Produktivität: Die Mythen der 'Industrie 4.0'. *Herder Korrespondenz, 9*, 32–35.

Gheaus, A., & Herzog, L. (2016). The goods of work (other than money!). *Journal of Social Philosophy, 47*(1), 70–89. https://doi.org/10.1111/josp.12140

Kirchschlaeger, P. G. (2023). Was sind Identität(en)? Eine menschenrechtsethische Betrachtung. In P. G. Kirchschlaeger (Ed.), *Menschenrechte: Aktuelle philosophische und ethische Fragestellungen* (Religionsrechtliche Studien (Vol. 6), pp. 51–74). TVZ Theologischer Verlag Zürich.

CHAPTER 7

Unconditional Basic Income

Abstract In this chapter, the unconditional basic income-approach is taken into consideration from an ethical standpoint. It contains the following core elements: (a) a monetary transfer, the amount of which secures the existence of the individual and enables him or her to take part in society. (b) No means-testing will take place prior to the payment (social-administrative test of income and assets) and without being forced to work for money or contribute in another way. From an ethical perspective, however, the unconditional basic income is not enough. This becomes evident because an unconditional basic income cannot cover all the previous 40 functions of a paid job.

Keywords Unconditional basic income • Ethical standpoint • Monetary transfer • Securement of existence • Participation in society • Means-testing • Salary • Functions of paid professional tasks

Decoupling income from work, the concept of the universal basic income or unconditional basic income is considered a possibility to react to the massive reduction of paid professional tasks for humans (Ford, 2015; Walker, 2014). The unconditional basic income-approach knows various forms and models. They all seem to contain the following core elements

P. G. Kirchschlaeger, *Ethics and the Digital Transformation of Human Work*, https://doi.org/10.1007/978-3-031-98704-5_7

that could shape a definition of an unconditional basic income: (a) The unconditional basic income contains a monetary transfer, the amount of which secures the existence of the individual and enables him or her to take part in society. (b) No means-testing will take place prior to the payment (social-administrative test of income and assets) and without being forced to work for money or contribute in another way. This means that the unconditional basic income secures a livelihood, enables participation in society, will be paid permanently and independent of needs. The unconditional basic income can be classified as a transfer payment that is independent of income and assets and is paid without means- or asset-testing (Schneider & Dreer, 2017).

From an ethical perspective, informed by the ethics of human rights protecting elements and areas of human existence that humans need to survive and live as humans, i.e., in human dignity, it can be deduced that with the approach of the unconditional basic income, the survival of all humans will be guaranteed by providing financial security to all humans, which is positive. At the same time, a significant contribution is made to social peace.

Human rights as an ethical frame of reference, however, also suggest that the unconditional basic income is not enough (Bothfeld, 2018; De Stefano, 2018). Not only the physical survival of humans is to be considered, but a life as a human—a life with human dignity. By enabling all humans to participate in society, a life as humans—a life with human dignity—for all humans is addressed. However, the unconditional basic income seems to be limited to one certain aspect of human life and a life with dignity. This becomes particularly obvious, if one reflects upon the possible 40 functions of paid professional tasks. It becomes evident that an unconditional basic income cannot cover all the previous functions of a paid job, which leads to the question about whether a possible adjustment of the unconditional basic income could fulfill some—or even all—of those 40 functions. Even if compensated by some money, freedom from work is not the freedom everybody wants (Santoni de Sio, 2024b, p. 26).

This question gains even more importance, on the one hand, because of the side effects and collateral phenomena of digital transformation and DS, which can be summarized in tendencies towards loneliness, individual isolation, seclusion from other people and from the society, degeneration of the ability for human interaction, and the development of a harmful and self-harmful self-image.

On the other hand, discussing this question gains importance when thinking that there will never be a shortage of tasks that are important for the whole society—in particular, tasks that are based on human interaction and relationships—on the contrary, they might even massively increase. For example, the demographic development and its acceleration based on medical advances will lead to the situation in which more generations of humans will live at the same time, so the significance of furthering intergenerational understanding will grow. At the same time, one must acknowledge that these tasks for the good of a society as a whole will probably neither be marketable nor be profitable.

Beyond that,

> Given what we know about how inequality and democracy interact, interventions should focus either on reducing AI-driven inequality or on directly strengthening democracy to safeguard against the fallout from increased disparities in income, wealth, and power. Because AI advances are likely to worsen inequality substantially, it will require a combination of these strategies to forestall the worst effects. (Bell & Korinek, 2023, p. 157)

Neither the reduction of inequality nor the promotion of democracy—although both serve the good of a society as a whole—will probably be marketable and profitable.

Moreover, "as effective as this may be in financially supporting unemployed people, the basic income strategy is silent about possible plans to prevent this structural mass unemployment from occurring in the first place" (Santoni de Sio, 2024a, pp. 4–5). This could be seen as a failure to live up to the human responsibility to shape and steer digital transformation and DS in an ethically positive way.

Looking at digital transformation and DS—in particular, their effects on the world of human work—and the idea of an unconditional basic income from an ethical perspective informed by the ethics of human rights, this analysis results in identifying ethical challenges. They occur in areas that are relevant to ethics of human rights, in digital transformation and DS, and their impact on the world of human work, as well as in respect to the unconditional basic income.

References

Bell, S. A., & Korinek, A. (2023). AI's economic peril. *Journal of Democracy, 34*(4), 151–161. https://doi.org/10.1353/jod.2023.a907696

Bothfeld, S. (2018). Das bedingungslose Grundeinkommen zwischen Utopie und sozialstaatlicher Wirklichkeit. *Leviathan, 46*(1), 81–108. https://doi.org/10.5771/0340-0425-2018-1-81

De Stefano, V. (2018). *Negotiating the algorithm: Automation, artificial intelligence and labour protection* (Employment Working Paper No. 246). International Labour Organization. Retrieved March 31, 2025, from https://www.ilo.org/publications/negotiating-algorithm-automation-artificial-intelligence-and-labour

Ford, M. (2015). *Rise of the robots: Technology and the threat of a jobless future.* Basic Books.

Santoni de Sio, F. (2024a). Artificial intelligence and the future of work: Mapping the ethical issues. *The Journal of Ethics, 28*(3), 407–427. https://doi.org/10.1007/s10892-024-09493-6

Santoni de Sio, F. (2024b). *Human freedom in the age of AI.* Routledge.

Schneider, F., & Dreer, E. (2017). *Grundeinkommen in Österreich?* Studie im Auftrag der initiative Wirtschaftsstandort Oberösterreich (IWS).

Walker, M. (2014). BIG and technological unemployment: Chicken litter versus the economists. *Journal of Evolution and Technology, 24*(1), 5–25. https://doi.org/10.55613/jeet.v24i1.10

CHAPTER 8

Society-, Entrepreneurship-, Research-Time-Model (SERT)

Abstract In this chapter, the Society-, Entrepreneurship-, Research-Time-Model (SERT) is introduced consisting of five pillars:

1. The payment of a higher basic income covering physical survival and a life with human dignity.
2. A conditional decoupling of income and work: Everyone contributes to the good of society ("society-time") in a freely chosen area.
3. The opportunity to enjoy during the "society-time" the 40 functions of paid professional tasks.
4. The incentives for entrepreneurship and for innovation: SERT comprises a reduction or exemption of "society-time" when committing to education, research, science, innovation, and entrepreneurship.
5. Distribution of growing economic profits: The profits from a more efficient and effective value-added chain are fairly distributed by harmonizing taxation globally, taxing capital, and taxing data-based systems (DS) and data.

Keywords Society-, Entrepreneurship-, Research-Time-Model (SERT) • Conditional basic income • "Society-time" • Functions of paid professional tasks • Incentives • Harmonization of taxation • Taxation of capital • Taxation of DS and data

P. G. Kirchschlaeger, *Ethics and the Digital Transformation of Human Work*, https://doi.org/10.1007/978-3-031-98704-5_8

The search for a legitimate alternative solution addressing the massive reduction of paid professional tasks for humans should—from an ethical perspective—consider the ethically positive potential of decoupling income and work but examine an alternative path to the unconditional basic income approach or universal income approach because of the ethical concerns amid the forty functions of paid professional tasks for humans mentioned above.

At the same time, it needs to consider the following:

> For policymakers working on labor market risks from general-purpose AI, key challenges include balancing AI innovation with worker protection, creating policy that can quickly adapt to evolving impacts, and ensuring economic benefits are shared both within and across countries. A central challenge is balancing innovation (which could boost productivity and growth) against protecting workers from wage declines and the harms associated with involuntary job loss […]: Policies must be adaptable given the speed of general-purpose AI development and the uncertainty about future impacts […]. However, policymakers face the challenge of setting flexible policy while still providing enough regulatory certainty to facilitate business investments and worker training decisions. Closely monitoring key trends can help policymakers increase their foresight into general-purpose AI's labor market impacts. These include sector-specific AI adoption rates, changes in wage distributions across industries, emergence of new job categories, and shifts in skill demands from employers as general-purpose AI systems progress and are more widely adopted. Finally, as the labor market impacts of AI are expected to vary significantly between countries, policymakers face challenges in coordinating international responses to prevent a widening global economic divide and ensure that AI can accelerate inclusive economic growth. (AI Action Summit, 2025, p. 118)

Besides moving away from the unconditional basic income approach or universal income approach, further aspects to adhere to encompass the balance between innovation and ethics of human rights as well as justice, flexibility, and adaptability while simultaneously maintaining and promoting legal security and certainty constituting a business- and innovation-friendly environment, as well as, last but not least, fostering inclusivity.

A different road to the unconditional basic income approach or universal income approach builds the change from an unconditional decoupling to a decoupling of income and work *without unconditionality* recognizing the ethical relevance of the forty functions of paid professional tasks for humans. From an ethical perspective, this could make a difference because one would expect an impact of the omission of the unconditionality on the potentially perpetual experience of the forty functions of paid professional tasks for humans as possible conditions for a basic or universal income could embrace the fulfillment of some if not all forty functions of paid professional tasks for humans including the corresponding experiences for humans. In other words, setting conditions for the decoupling of income and work would create the opportunity to include in these conditions some if not all up to now existing forty functions of paid professional tasks for humans.

The *Society-, Entrepreneurship-, Research-Time-Model (SERT)* (Kirchschlaeger, 2019a, 2021) represents a decoupling of income and work *without unconditionality* aiming at contributing to the survival of all humans and to life as humans—a life with human dignity. SERT encompasses five pillars:

1. On the one hand, it is based on the *payment of a higher basic income*, which will not only cover the needs for a physical survival and existence of humans but also allow for a life as a human—a life with human dignity.
2. On the other hand—and here the unconditional side of the basic income will be substantially adapted to a *conditional decoupling of income and work*—it consists of a commitment of the same time-duration by each person to society (society-time), which is demanded of each person in return for the basic income. Analogously to the already since almost three decades existing, tried and tested, and positively proven model of the Swiss civil service which represents an alternative to the Swiss military service, every person could choose from a broad array of options to contribute to the good of society as a whole in one chosen area. Digital transformation of human labor and the impact of DS on human work will not make these professional tasks for the good of society as a whole for humans disappear. But digital transformation and DS will either make the willingness of the private sector to pay any humans for fulfilling them disappear. Or digital transformation and DS will not increase the already low or nonexistent willingness of the private sector to address these for the good of society as a whole at all (due to the fact that these professional tasks are not profitable, taking into account the fact that not every aspect of human existence and of societal coexistence is

marketable; e.g., assisting, accompanying, or taking care of elderly humans who need support is not a profitable business but the right thing to do from an ethical perspective based on the principle of inclusion and the principle of non-discrimination). SERT embraces these professional tasks for the good of society as a whole for humans with its pillar "society-time."

The time-duration of the "society-time," in other words for this commitment to society at large, would be the same for everyone. From an ethics of human rights perspective, the same time-duration for everyone as well as the self-determination of each individual to choose the area of one's contribution for the good of society as a whole are essential in order to pay the highest possible respect to the core principles of human rights, namely, human dignity, equality, freedom, and autonomy of all humans.

This free and autonomous choice of one's own individual contribution as a duty corresponding to the coverage of the costs for existence and for a life as a human—a life with human dignity—would allow to realize the freedom and autonomy of every individual in this model of decoupling income from work.

3. SERT—and this is the main ethical reason for this model—would offer all humans the opportunity to enjoy during the "society-time" *some or all forty functions of paid professional tasks.*

   Furthermore, SERT would address the above-mentioned ethical challenges.

4. Beyond that, SERT would keep the *incentives for entrepreneurship and for innovation* alive and avoid human discouragement. The former would fall away with an unconditional basic income or universal basic income approach. Incentives for entrepreneurship and for innovation are not only of economic significance but also of ethical relevance because they are essential for progress in a society: The economic system of a social market economy is characterized, on the one hand, by its principles of a social network based on solidarity and justice. On the other hand, the principles of fairness and free competition, as well as the logic of the market, offer incentives for education, science and research, innovation, and entrepreneurship. SERT takes this into account so as to encourage the striving for those objectives by offering a reduction in the time-duration of "society-time" or even an exemption from "society-time" when

committing oneself to education, research and science, innovation, and entrepreneurship. This way, incentives for education, research and science, innovation, and entrepreneurship can be given in order to ensure that education, research and science, innovation, and entrepreneurship remain an attractive option.

5. *Distribution of growing economic and financial profits:* Addressing concretely the core consequence of the digital transformation of society and economy and the use of DS mentioned above—namely, ***ever fewer humans are directly involved economically and socially in a more efficient and more effective value-added chain***—is of ethical relevance in order to avoid a widening of local, national, regional, and global inequality and injustice, in order to open equal opportunities for everyone, and in order to guarantee peaceful coexistence in a just society. "The share of wages in economic output will continue to fall as digitization progresses, while the share of capital will increase. […] More than ever before, money will be earned with money, not with work. Those who work are the new poor" (Hofstetter, 2017, p. 88, translated by the author).

   The amplification of domains mastered by DS intensifies the ethical challenges provoked by the core consequence of the digital transformation of society and economy and the use of DS. "General-purpose AI-driven labor automation is likely to exacerbate inequality by reducing the share of all income that goes to workers relative to capital owners" (AI Action Summit, 2025, p. 115).

   The good news: A more efficient and effective value-added chain results in more profits and resources at the end so a higher basic income not only guaranteeing physical survival for all humans but also a life with human dignity—a life as a human—does not fail because of a lack of financial means. The bad news:

> If human labor is indeed rendered superfluous by automation, then our chief economic problem will be one of distribution, not of scarcity. The primary system of income distribution in market economies is rooted in labor scarcity; citizens possess (or acquire) a bundle of valuable 'human capital' that, due to its scarcity, generates a flow of income over the career path. If machines were in fact to make human labor superfluous, we would have vast aggregate wealth but a serious challenge in determining who owns it and how to share it. One might presume that with so much wealth at hand, distribution would be relatively straightforward to resolve. But history suggests that this

> prediction never holds true. There is always perceived scarcity and ongoing conflict over distribution, and I do not expect that this problem will become any less severe as automation advances. Are we actually on the verge of throwing off the yoke of scarcity so that our primary economic challenge soon becomes one of distribution? (Autor, 2015, p. 28)

More profits and resources stemming from a more efficient and effective value-added chain need to be divided more actively as the massive reduction of paid professional tasks does exclude most people from participating in, contributing to, as well as partaking in the value creation.

One way to address these issues is to transform the economic non-involvement and the concentration of the power of DS in the hands of a few multinational technology-companies (Wong, 2024) to a participatory model (Kochan et al., 2024) and to a functioning and efficient liberal market.

> There is only one solution to the long-term challenge posed by machines substituting for human skills and reducing demand for skilled labor. That is for you, me, all of us to have a substantial ownership stake in the robot machines that will compete with us for our jobs and be the vehicle for capital's share of production. We must earn a substantial part of our incomes from capital ownership rather than from working. Unless workers earn income from capital as well as from labor, the trend toward a more unequal income distribution is likely to continue, and the world will increasingly turn into a new form of economic feudalism. We have to widen the ownership of business capital if we hope to prevent such a polarization of our economies. There are diverse pathways to spread the ownership of capital. Ownership can take the form of worker assets in private pension funds or other collective savings vehicles that invest in shares on the stock market or that invest directly in equity in other firms. It can also take the form of workers buying shares or putting money in mutual funds themselves. But the form of ownership that potentially has the greatest economic benefit in dealing with robotization and the falling share of labor income is employee ownership. (Freeman, 2015)

SERT foresees that promoting and fostering the widening of the ownership of business capital in these more efficient and effective value creation-processes allows a broader partaking.

Beyond that, the common and proven way to cover public services, societal needs, and government expenditures, and by this, to indirectly—especially in a progressive tax-system—distribute income and wealth to balance inequalities forms taxes. This means that—besides the widening of the ownership of business capital—SERT consists also in an adaptation of the tax-system into a fair global tax system (Mehta et al., 2020; Pogge & Mehta, 2016), taking into account the concentration of more income and wealth in less hands as an effect of massive reduction of paid professional tasks for humans due to the digital transformation of the paid labor-market and the impact of the use of DS on human work.

Why taxes?

> Tax justice matters because a modern economy requires that the state has sufficient revenue over time to fund the physical and social infrastructure essential to economic welfare. It should also enable a degree of wealth redistribution between rich and poor people to promote equity and security. Failure to achieve the first objective because of poorly designed, unfair or leaking tax systems is likely to yield economic failure, while failure to achieve the second objective will lead to social failure. In either case, the costs to society are enormous. Tax justice is, therefore, at the heart of stable and democratic forms of government. (Tax Justice Network, 2005, p. 11)

This adaptation of the tax-system should include harmonizing taxation globally, taxing capital rather than work, taxing DS, and taxing data.

First steps in the right direction have been made regarding the harmonization of taxation globally considering an observation by at that time German Finance Minister Olaf Scholz together with at that time Spain's Minister of Economics Nadia Calvino, at that time Italy's Minister of Finance Roberto Gualtieri, and at that time French Finance Minister Bruno Le Maire, namely that "these (digitally-based companies) generate enormous profits in regions where they have no official base, and therefore do not pay taxes. At the same time, some of the world's largest corporations continue to direct their profits to countries with lower tax rates" (Deutsche Welle, 2020; Scholz et al., 2020). The basis for these efforts to harmonize taxation globally represents the OECD/G20 Inclusive Framework on base erosion and profit shifting (BEPS). BEPS refers to tax planning strategies that exploit gaps and mismatches in tax rules to artificially shift profits to low or no-tax locations where there is little or no economic activity or to erode tax bases through deductible payments such as interest or royalties. Although some of the schemes used are illegal, most are not. This undermines the fairness and integrity of tax systems because businesses that operate across borders can use BEPS to gain a competitive advantage over enterprises that operate at a domestic level. Moreover, when taxpayers see multinational corporations legally avoiding income tax, it undermines voluntary compliance by all taxpayers. Developing countries' higher reliance on corporate income tax means they suffer from BEPS disproportionately. BEPS practices cost countries USD 100–240 billion in lost revenue annually. Working together within OECD/G20 Inclusive Framework on BEPS, over 135 countries and jurisdictions are collaborating on the implementation of 15 measures to tackle tax avoidance, improve the coherence of international tax rules, and ensure a more transparent tax environment (Organisation for Economic Co-operation and Development [OECD], n.d.-b).

The Inclusive Framework on BEPS employs a two-pillar approach consisting in (first pillar: unified approach) new rules regarding where tax should be paid ("nexus" rules), and in new rules regarding what portion of their profits should be taxed ("profit allocation" rules) ("The aim is to ensure that multinational enterprises (MNEs) conducting sustained and significant business in places where they may not have a physical presence can be taxed in such jurisdictions" (OECD, 2020)), as well as consisting of (second pillar) tackling remaining base erosion and profit shifting (BEPS) issues, and ensuring that international businesses pay a minimum level of tax (OECD, 2020). In 2021, the Global Minimum Tax was introduced and started to apply at the beginning of 2024.

> The global minimum tax, which is based on the Global Anti-Base Erosion (GloBE) Model Rules, ensures that large multinational enterprises pay a minimum level of tax on their income in each jurisdiction where they operate, thereby reducing the incentive for profit shifting and placing a floor under tax competition, bringing an end to the race to the bottom on corporate tax rates. (OECD, n.d.-a)

This means that a top-up tax is imposed on multinational corporations enjoying on a jurisdictional basis an effective tax rate that is below the minimum rate of 15% (OECD, n.d.-a). The Global Minimum Tax ensures

> greater fairness in the allocation of taxing rights and tax revenue among the world's countries. Such rules are particularly important for the taxation of

> large tech companies – for example, companies that can reap extremely high profits from internet sales or ad clicks even in countries where they have no factories or branches. Under the current rules, taxes are payable primarily in the countries where firms have a physical presence. We need new rules that adequately reflect the increasing digitalization of the economy. Companies should also pay taxes in the countries where they generate profits. The new approach will also ensure that all companies make a fair contribution to financing public goods. (Federal Ministry of Finance, 2024)

Regarding taxing capital rather than work, it needs to be perceived that

> capital is not a thing [...]. Instead, capital is often depicted as a social relation that has been forged in the production process between capitalists. Missing in this social relation is the critical role of the state and of law. Capital is coded in law [...]. It piggybacks on and at times hijacks the state's institutional powers of coercion to create wealth for holders of capital assets. Four critical attributes that are fashioned in law distinguish capital assets from any other asset: the holders of capital have priority rights over competing claimants; they can extend their rights in time, making them durable; further, the holders may be given the right to convert these assets into safe assets that can retain their (nominal) value. Last but not least, universality ensures that the above attributes, priority, durability, and convertibility can be enforced against the world. They bind not only the parties to a transaction but anyone, because they are backed by state power. (Pistor, 2019, p. 430)

The understanding of these characteristics of capital helps the taxation of capital with legal means as it opens direct access to its fundament, namely being coded in law. At the same time, the particular role states played so far needs to be taken into account as well.

> The fragile balance between collective governance and private autonomy, both of which are guaranteed by law, is difficult to sustain even within a single polity like a nation-state. [...] Globalization has made this balance even more fragile. In the hope of promoting economic development, states have made legal devices portable. They have allowed private actors not only to opt out of their own legal system but to opt into other legal systems of their choice, without compromising their access to the coercive law enforcement institutions that states command. They have empowered private actors to defend their interests against themselves, not only in their own courts but even in offshore arbitration tribunals. In short, states have become entan-

> gled by a web of legal claims that they helped to create, which have weakened their ability to use law as a means for collective self-governance. States have empowered capital to rule by law and in the course of this they have allowed themselves to be ruled by capital. (Pistor, 2019, p. 439)

At the same time, it must be admitted that taxing capital rather than work is not a new idea (Feige, 1989; Keynes, 1936; Summers & Summers, 1989; Tobin, 1978), but the taxation of financial transactions gains importance and relevance due the following impact of digital transformation and the use of DS: Regarding taxing DS, and taxing data,

> the profits of the large digital groups – be they American, European or Chinese – are not taxed appropriately. These companies make huge profits in areas where they have little physical presence but exploit the data of millions of users. Often the companies with the highest profits end up paying the lowest taxes. As a result, they do not make a fair contribution to the financing of our community. This situation is unacceptable, inefficient and above all unsustainable. With international digital taxation, we can eliminate this problem by establishing fair and stable taxation that is appropriate to the new business models of the digital economy. At the same time, it will give the companies concerned greater fiscal legal certainty. (Scholz et al., 2020)

More profits, more financial means, more resources, more income, more wealth, all in fewer hands of fewer people, constitute the reasons why SERT—in order to contribute to the realization of human dignity and human rights for all humans and to a peaceful coexistence in a society and in order to prevent injustice, inequality, and social unrest—envisages a distribution of the economic and financial success of digital transformation and the use of DS by widening the partaking in business capital, as well as the adaptation of taxation into a fair global tax-system taking into account the technology-based change.

Considering potential counterarguments against SERT in general, one conceptional opposition could grow out of some worries about the potential high administrative burden to set up and run SERT. The administration of SERT could be addressed with a digital answer. All supervising, reporting, and controlling processes can be digitalized. Of course, solutions based on blockchain technology (Kirchschlaeger, 2023) could contribute to these efforts.

Potential counterarguments against SERT in general could consist in criticism nourished by worries about abuse of SERT—e.g., some people could try not to serve "society-time"—which can be addressed with the likely realistic assessment that this system, as any other system, will recognize abuse attempts. As long as this concerns only a small, statistically irrelevant number, which is to be expected because of the inherent motivation of humans to the good of a society as a whole as well as because of the digital supervising, reporting, and controlling processes, there is no need to worry. This is particularly true because it also needs to be considered that even in case of abuse, the commitment and concrete service to society and to the good of the society as a whole would be much greater than in the case of the unconditional basic income or without any decoupling of income from work.

Another potential counterargument against SERT in general could be that SERT is not affordable because a basic income in general and especially a higher basic income as in the case of SERT guaranteeing not only physical survival but also a life as a human—a life with human dignity—would blast the economic and financial possibilities of an economy and a society. Taking the core consequence of digital transformation and the use of DS seriously, namely that *ever fewer humans are directly involved economically and socially in a more efficient and more effective value-added chain* (Kirchschlaeger, 2016, 2017b), means that this counterargument can be refuted because more value will be created.

> The basic economics of automation involves technologically advanced machines replacing workers in the production process. The result is a rise of output accompanied by a fall in the demand for labor. Output rises, labor productivity rises, but wages fall, and income is redistributed from labor to capital. Is society better off? Yes and no. Gross domestic product rises, but workers are left poorer. Thus, the way to ensure that everybody benefits from the technological advances brought by automation is to accompany the shift with a transfer of income from capital (the automation 'winner') to workers (the automation 'loser'). (Sachs, 2019, p. 159)

The main challenge from an ethical perspective is that fewer humans contribute directly to the value-added chain and benefit from this added value: Not a lack of financial means turns out to be the issue because more efficient and more productive value-added chains lead to an increase in that regard, but rather the question of justice and human rights. At stake is the distribution of the added value, which is created. At the center of concerns is also the question of social integration as fewer humans are involved in the value-added chain based on digital transformation of human work and the impact of the use of DS on human labor—a worry that represents the core consequence of digital transformation and the use of data-based systems in general (Kirchschlaeger, 2019b). "It's one of the dirty secrets of economics: technology progress does grow the economy and create wealth, but there is no economic law that says everyone will benefit" (Rotman, 2013).

SERT would be part of an effort to tackle this challenge, which is part of a broader problem, namely that, especially on a global level, digital transformation and DS run the risk of serving as a vehicle contributing to widening the gap between rich and poor. This is happening due to the lack of access to digital transformation and DS and would strengthen already established privileged positions. The "digital divide" is and—if severe counteractions are not taken—remains an unjust reality. Its global solution must be approached and considered as a priority today and tomorrow.

> People who have access to the digital world have so far benefited economically from this process in comparison to people who do not have access ('digital divide'). However, in the future almost everyone will have access to the digital world, but with different quality characteristics in terms of access. There is a tendency to create a state of 'digital inequality', similar to the economic inequality that has recently received more media attention. This inequality needs to be overcome through inclusive global governance with a balance between central and distributed control. (Uenver, 2015, p. vi)

From an ethical perspective informed by human rights, the digital gap-widening-impact of digital transformation and DS urge humans to act rapidly insofar as digital transformation and DS further fuel a problem that already existed before. Slavery-like working conditions or modern slavery used for the extraction of raw materials (Amnesty International, 2016; Gerding, 2019; Odiot & Gallet, 2012; Schweizer, 2015), and the production of technology products as well as huge differences in political participation for shaping digital transformation and the use of DS, as well as with regard to enjoying the sunny side (*inter alia* equal opportunities, profit-sharing) or suffering from the negative sides of this transformation (*inter alia* exclusion, exploitation that is degrading and violates human rights), cry for fundamental change and ethical progress. This is not a new challenge, but one which is still topical, and which is growing rapidly as a result of technology-based progress. An increase in the demand for these natural resources in digital transformation and DS worsens these human rights violations. The increased demand for technology products underlines the urgency to establish humane working conditions in all phases of value creation, including production. The increased demand for natural resources fueling digital transformation and DS and for technology products calls for optimizing the implementation of already existing human rights-obligations of states and the private sector in this area (Kirchschlaeger, 2015, p. 280).

Of course, companies do make a social contribution when trying to make profits and grow. Because, apart from the immediate economic effects on society (e.g., taxes), the positive, value-adding, and solution-oriented power of entrepreneurship for the context it operates in also needs to be considered. Moreover, its dynamic inspires society, as is becoming so evident in the field of digital transformation and DS. People and society benefit from this innovation potential of profit-oriented decision-making and acting, because economic innovations trigger or drive social progress. Furthermore, the business approach is used to address questions, to solve problems, and to meet challenges, because those solutions are expected to be more sustainable than philanthropic or solidarity-based solutions.

In this sense, social entrepreneurship highlights an important quality of entrepreneurship in general: the connection between economic and ethical rationality. Companies do not only create economic benefit but also contribute socially. Companies contribute to social systems. "The central role of business (corporate management) is in ensuring the long-term viability of a democratically governed society [...] grounded in justice, equality, and trust and supported by sustainable natural, social, and economic systems" (Dillard, 2013, p. 199). At the same time, societies give companies the necessary scope for decision-making and action.

> Again and again, the interaction between companies and society comes to light: the society needs strong companies to ensure its well-being. Companies, on the other hand, need strong societies. Because only in a strong society they find stable political environments, resources and markets, infrastructure, intellectual and financial capital, good workers, partners, and customers. (Clever & Ramb, 2010, p. 59)

However, there is the danger that by putting more importance on economic rationality, the vision of companies will be narrowed and that they might repress ethical rationality and alienate themselves from their social contribution. For example, companies can be involved in exploitation and suppression or observe them indifferently, which unfortunately happens in particular in the field of digital transformation and the use of DS. In both cases, the following applies: If the above-mentioned social contributions of companies are connected to such practices, they automatically lose their value for society due to the antisocial nature of this business practice. Furthermore, they are an injustice and a threat to society. An imbalance between economic and ethical reality can be ascertained. This imbalance leads to the fact that the positive impact created by companies will not only be eradicated, but only the harm done to society will count. For example, inhumane working conditions would negate the social contribution of that business and turn the acting company into a criminal and a violator of human rights. The argument that this creates, e.g., tax-revenues, loses all relevance because of the injustice of inhumane working conditions. Not only that, but the company will also be legitimately criticized for it and would—from an ethical point of view—also lose its raison d'être.

The interaction between economic and ethical rationality in business actions is important for the whole society. Societies depend on companies taking on their social responsibilities, because companies have power and influence. Companies not only have the power to influence (Ekmekçi, 2014) the behavior of their investors (Gurbuz et al., 2014; Idowu & Yuksel Mermod, 2014), business partners, suppliers, and customers (Schoenheit, 2014, pp. 41–48). The intensity of the mutual interdependence in the course of globalization multiplies the power of companies, which means a society cannot cope with a failure or partial failure of companies providing a contribution to society as a whole or attacks on social cohesion by companies through exploitation and suppression (Dillard & Murray, 2013, p. 11). "One instrument brought into play by companies to achieve more economic fairness is the so-called Corporate Social Responsibility (CSR). Those are standards that companies gave themselves with regards to their business conduct (e.g., working conditions, environmental behavior, social aspects)" (Hilty & Henning-Bodewig, 2014, p. 4).

Corporate Social Responsibility (CSR) as a concept widens the objects of responsibility for companies beyond the shareholders and includes the responsibility of companies "towards internal and external stakeholders, employees, customers, suppliers and the closer and broader community" (Braun, 2009, p. 249). It is based on the term "responsibility" (Kirchschlaeger, 2014).

> The total corporate social responsibility of business entails the simultaneous fulfillment of the firm's economic, legal, ethical, and philanthropic responsibilities. Stated in more pragmatic and managerial terms, the CSR firm should strive to make a profit, obey the law, be ethical, and be a good corporate citizen. (Carroll, 1991, p. 43)

The private sector needs to do its bit toward a humane and sustainable future for humanity and the planet by substantially improving its human rights performance. Besides these corporate human rights obligations (Kirchschlaeger, 2017a), it is part of the human rights obligations of states to make sure that corporations do not violate human rights and that human rights are respected, protected, implemented, and realized in the economic sphere (Kirchschlaeger, 2015).

SERT attempts to address the technology-based massive reduction of paid professional tasks for humans in a way which upholds human rights of all humans.

> This complex problem may require a complex response through a collective effort of all the parties involved. Among other things, we need State's policies to support (temporary) unemployed people, regulation of gross violations of fundamental rights such as privacy infringements and physical and psychological harm, and general protection of decent work conditions. We need a broader policy and regulatory approach to govern new forms of precarious labor, such as the gig economy, and prevent new exploitation and injustice. We need to discuss and control the power of the big tech companies in the light of their impact on the environment, global justice and people's moral and political freedom. Employers must become more aware of their power to shape human-technology relation and the ethical and economic value of promoting meaningful work. From a research point of view, universities and research institutes should create and support long-term multidisciplinary research programs connecting normative (philosophical, political), empirical, and technical perspectives to jointly identify the dimensions of good work in different domains, in collaboration with workers and employers, as well as new design methodologies. These programs should eventually prototype and test AI systems that promote rather than harm ethical and societal values. (Santoni de Sio, 2024, p. 423)

Maybe SERT as a model itself and with its corresponding questions can be used as an ethical impulse for continued interdisciplinary discourse.

Due to the foreseeable ethical ugliness of a transitional phase characterized by massive reduction of paid professional tasks and unprepared society and economy still pretending to strive for full employment, which comes closer and closer with immense speed because of the rapid and exploding technological progress in the area of digital transformation and DS, humans should not lose time to start addressing the end of the striving for full employment and to start implementing SERT. "The progress of information technology continues to accelerate and the impact on the job market could well materialize long before we are prepared for it" (Ford, 2018, p. 45).

References

AI Action Summit. (2025, January). *International AI safety report. The international scientific report on the safety of advanced AI.* Retrieved March 31, 2025, from https://www.gov.uk/government/publications/international-ai-safety-report-2025

Amnesty International. (2016, January 19). *Kinderarbeit für Mobiltelefone und Elektroautos.* Medienmitteilung. Retrieved March 31, 2025, from https://www.amnesty.ch/de/laender/afrika/demokr-rep-kongo/dok/2016/bericht-kinderarbeit-fuer-mobiltelefone-und-elektroautos

Autor, D. H. (2015). Why are there still so many jobs? The history and future of workplace automation. *The Journal of Economic Perspectives, 29*(3), 3–30. https://doi.org/10.1257/jep.29.3.3

Braun, H. (2009). Verantwortung. *Die Neue Ordnung, 63*(4), 244–252.

Carroll, A. B. (1991). The pyramid of corporate social responsibility: Toward the moral management of organizational stakeholders. *Business Horizons, 34*(4), 39–48.

Clever, P., & Ramb, C. (2010). Freiheit ohne Verantwortung verkommt. Die Finanz- und Wirtschaftskrise – Folgen für unternehmerisches Handeln und die gesellschaftliche Verantwortung von Unternehmen. In M. Heimbach-Steins (Ed.), *Weltwirtschaft und Gemeinwohl. Eine Zwischenbilanz der Wirtschaftskrise* (pp. 53–63). Aschendorff.

Deutsche Welle (DW). (2020, February 22). No time to wait' for tax reforms, says German finance minister. *Deutsche Welle.* Retrieved March 31, 2025, from https://www.dw.com/en/no-time-to-wait-for-tax-reforms-says-german-finance-minister/a-52475061

Dillard, J. (2013). Human rights within an ethic of accountability. In K. Haynes, A. Murray, & J. Dillard (Eds.), *Corporate Social Responsibility. A research handbook* (pp. 196–220). Routledge.

Dillard, J., & Murray, A. (2013). Deciphering the domain of Corporate Social Responsibility. In K. Haynes, J. Dillard, & A. Murray (Eds.), *Corporate Social Responsibility. A research handbook* (pp. 10–27). Routledge.

Ekmekçi, A. (2014). An examination of the relationship between companies' Corporate Social Responsibility (CSR): Activities and consumers' purchase behavior. In A. Yuksel Mermod & S. O. Idowu (Eds.), *Corporate Social Responsibility in the global business world* (pp. 49–73). Springer.

Federal Ministry of Finance. (2024, January 12). *Global minimum tax: Frequently asked questions.* Federal Ministry of Finance Resources, FAQs. Retrieved March 31, 2025, from https://www.bundesfinanzministerium.de/Content/EN/FAQ/faq-global-minimum-tax.html

Feige, E. L. (1989, August 28). *Taxing all transactions: The automated payment transactions (APT) Tax system.* International Institute of Public Finance Conference on Public Finance and Steady Economic Growth. Buenos Aires, Argentina

Ford, M. (2018). The rise of the robots: Impact on unemployment and inequality. In E. Paus (Ed.), *Confronting dystopia: The new technological revolution and the future of work* (pp. 27–45). Cornell University Press. https://doi.org/10.7591/9781501719868

Freeman, R. (2015). Who owns the robots rules the world. *IZA World of Labor.* https://doi.org/10.15185/izawol.5

Gerding, J. (2019, July 18). Ein Rohstoff und sein Preis. *Die Zeit.* Retrieved March 31, 2025, from https://www.zeit.de/2019/30/kobalt-kongo-rohstoff-elektroautos-smartphones-bergbau

Gurbuz, A. O., Gokmen, M. K., & Aybars, A. (2014). Do institutional investors prefer to invest in socially responsible companies? An empirical analysis in Turkey. In A. Yuksel Mermod & S. O. Idowu (Eds.), *Corporate Social Responsibility in the global business world* (pp. 311–324). Springer.

Hilty, R., & Henning-Bodewig, F. (2014). Vorwort und Einführung in die Thematik. In R. M. Hilty & F. Henning-Bodewig (Eds.), *Corporate Social Responsibility – Verbindliche Standards des Wettbewerbsrechts? MPI studies on intellectual property and competition law 21* (pp. 3–5). Springer.

Hofstetter, Y. (2017). EMANZIPIERT EUCH! Menschwerdung im digitalen Zeitalter. In B. Spiess & N. Fabisch (Eds.), *CSR und neue Arbeitswelten: Perspektivwechsel in Zeiten von Nachhaltigkeit, Digitalisierung und Industrie 4.0* (pp. 73–90). Management-Reihe Corporate Social Responsibility. Springer.

Idowu, S., & Yuksel Mermod, A. (2014). Investing peacefully. A global overview of socially responsible investing. In A. Yuksel Mermod & S. O. Idowu (Eds.), *Corporate Social Responsibility in the global business world* (pp. 325–355). Springer.

Keynes, J. M. (1936). *The general theory of employment, interest and money.* Macmillan Cambridge University Press.

Kirchschlaeger, P. G. (2014). Verantwortung aus christlich-sozialethischer Perspektive. *ETHICA, 22*(1), 29–54.

Kirchschlaeger, P. G. (2015). Multinationale Konzerne und Menschenrechte. *ETHICA, 23*(3), 261–280.

Kirchschlaeger, P. G. (2016, March 30). Digitalisierung und Robotisierung der Gesellschaft aus theologisch-ethischer Perspektive. Retrieved March 31, 2025, from *Feinschwarz.net*. https://www.feinschwarz.net/digitalisierung-und-robotisierung-der-gesellschaft-aus-theologisch-ethischer-perspektive/

Kirchschlaeger, P. G. (Ed). (2017a). *Die Verantwortung von nichtstaatlichen Akteuren gegenüber den Menschenrechten* (Religionsrechtliche Studien (Vol. 4)). TVZ Theologischer Verlag Zürich.

Kirchschlaeger, P. G. (2017b). Roboter und Ethik. *Aktuelle Juristische Praxis, 26*(2), 240–249.

Kirchschlaeger, P. G. (2019a). Bedingungsloses Grundeinkommen – eine menschenrechtsethische Betrachtung. In P. Hladschik & F. Steinert (Eds.), *Menschenrechten Gestalt und Wirksamkeit verleihen: Making Human Rights Work. Festschrift für Hannes Tretter und Manfred Nowak* (pp. 551–563). NWV.

Kirchschlaeger, P. G. (2019b). Digital Transformation of Society and Economy. Ethical Considerations from a Human Rights Perspective. *International Journal of Human Rights and Constitutional Studies, 6*(4), 301–321.

Kirchschlaeger, P. G. (2021). *Digital Transformation and Ethics. Ethical Considerations on the Robotization and Automation of Society and the Economy and the Use of Artificial Intelligence*. Nomos.

Kirchschlaeger, P. G. (2023). Blockchain Ethics. *Philosophies, 9*(1), 2. https://doi.org/10.3390/philosophies9010002

Kochan, T. A., Armstrong, B., Shah, J., Castilla, E. J., Likis, B., & Mangelsdorf, M. E. (2024). Bringing worker voice into generative AI. *An MIT Exploration of Generative AI*. https://doi.org/10.21428/e4baedd9.0d255ab6

Mehta, K., Shubert, E., & Dayle Siu, E. (2020). *Tax justice and global inequality*. Zed Books Ltd.

Odiot, A., & Gallet, A. (2012). *Sambia: Wer profitiert vom Kupfer?* [Video; no longer available]. YouTube.

OECD. (2020, February 13). *OECD presents analysis showing significant impact of proposed international tax reforms*. OECD Web archive. Retrieved March 31, 2025, from https://web-archive.oecd.org/2020-03-03/545463-oecd-presents-analysis-showing-significant-impact-of-proposed-international-tax-reforms.htm

OECD. (n.d.-a). *Global minimum tax*. OECD Topics. Retrieved March 31, 2025, from https://www.oecd.org/en/topics/global-minimum-tax.html

OECD. (n.d.-b). *What is BEPS? Inclusive framework on base erosion and profit shifting*. OECD Topics. Retrieved March 31, 2025, from http://www.oecd.org/tax/beps/about/#mission-impact

Pistor, K. (2019). Capital's global rule. *Constellations, 26*(3), 430–441. https://doi.org/10.1111/1467-8675.12434

Pogge, T., & Mehta, K. (2016). *Global tax fairness.* Oxford University Press.

Rotman, D. (2013). How technology is destroying jobs. *MIT Technology Review, 16*(4), 28–35. Retrieved March 31, 2025, from https://www.technologyreview.com/s/515926/how-technology-is-destroying-jobs/

Sachs, J. D. (2019). Roundtable: AI and the future of global affairs. *Ethics & International Affairs, 33*(2), 159–167.

Santoni de Sio, F. (2024). Artificial intelligence and the future of work: Mapping the ethical issues. *The Journal of Ethics, 28*, 407–427. https://doi.org/10.1007/s10892-024-09493-6

Schoenheit, I. (2014). *Corporate Social Responsibility and consumers.* Springer.

Scholz, O., Calvino, N., Gualtieri, R., & Le Maire, B. (2020, February 20). Ein gerechtes internationales Steuersystem. *Die Welt.* https://www.welt.de/print/die_welt/debatte/article206059379/Gastkommentar-Ein-gerechtes-internationales-Steuersystem.html

Schweizer, D. (2015). *Dirty gold war – Dreckiges Gold.* Arte Doku [Documentary]. https://www.youtube.com/watch?v=WoK8cmdarIU

Summers, L., & Summers, V. P. (1989). When financial markets work too well: A cautious case for a securities transactions tax. *Journal of Financial Services Research, 3*(2–3), 261–286.

Tax Justice Network. (2005). *Tax us if you can: Why Africa should stand up for tax justice.* Tax Justice Network. Retrieved March 31, 2025, from https://www.taxjustice.net/cms/upload/pdf/tuiyc_-_eng_-_web_file.pdf

Tobin, J. (1978). A proposal for international monetary reform. *Eastern Economic Journal, 4*(3-4), 153–159. http://web.holycross.edu/RePEc/eej/Archive/Volume4/V4N3_4P153_159.pdf

Uenver, H. (2015). *Globale Vernetzung, Kommunikation und Kultur – Konflikt oder Konvergenz?* Dissertation, Universitaet Ulm.

Wong, W. K. O. (2024). The sudden disruptive rise of generative artificial intelligence? An evaluation of their impact on higher education and the global workplace. *Journal of Open Innovation Technology Market and Complexity, 10*(2), 100278. https://doi.org/10.1016/j.joitmc.2024.100278

CHAPTER 9

Outlook: Human Rights-Based Data-Based Systems (HRBDS) and an International Data-Based Systems Agency (IDA) at the UN

Abstract In this chapter, the Society-, Entrepreneurship-, Research-Time-Model (SERT) is embedded in a context which would be beneficial for the realization of SERT, consisting in human rights-based data-based systems (HRBDS) (a global regulatory framework encompassing the respect of human rights in the entire life-cycle of DS) and the establishment of an International Data-Based Systems Agency (IDA) at the UN promoting, enforcing, and realizing human rights-based DS by an access to market approval-process, by monitoring that human rights are not violated with or by DS, and by fostering international technical collaboration. Human rights-based DS and the establishment of IDA at the UN enjoy the support by, among others, UN Secretary General António Guterres, Pope Francis, His Holiness the Dalai Lama, Sam Altman (Founder of Open AI), and "The Elders."

Keywords Society-, Entrepreneurship-, Research-Time-Model (SERT) • Human rights-based data-based systems (HRBDS) • Global regulatory framework • Life-cycle • International Data-Based Systems Agency (IDA) at the UN • Access to market approval-process • Monitoring • International technical collaboration

P. G. Kirchschlaeger, *Ethics and the Digital Transformation of Human Work*, https://doi.org/10.1007/978-3-031-98704-5_9

9.1 Humans Should Shape Digital Transformation and DS

The creation of a fruitful normative ground and a supportive legal framework for SERT could be decisive for facilitating the implementation of SERT. Humans need to become active so that digital transformation and DS do not simply happen, but that humans shape them. This is necessary so that digital transformation and DS will not be reduced to an instrument serving pure efficiency but can rise to their ethically positive potential. More importantly, there is a need for ethical guidance to review the economic self-interests that run digital transformation and DS so far almost exclusively.

These efforts should channel into establishing global regulations and governance in the digital domain and in the sphere of DS to ensure that all humans and the planet share the ethical benefits of DS and that the ethical risks of DS are mastered. This is of essential significance not only because of, but especially in a time where multinational technology-companies who are the decisive driving-force of this technological progress officially and explicitly step down from their ethical responsibilities. For example, Meta announces that it will stop its ethically informed moderation of the content (e.g., fighting racist and sexist hate speech as well as incitements to violence) on their social media-platforms (Isaac & Schleifer, 2025). Google, for example, removes their "Responsible AI Principles" which it has so far used to engage in technologies violating human rights as well as in technologies for surveillance and for causing harm (e.g., DS for weapons) (Tiku & De Vynck, 2025). These are just additional pieces of explicit evidence (Zuboff, 2019) that self-regulation of multinational technology-companies neither will really happen nor work.

9.2 Existing Global Governance Initiatives

The debate about the international governance of DS (Kirchschlaeger, 2024a, 2024d, 2024e) encompasses several declarations, recommendations, principles, and guidelines—the first generation of governance initiatives: "the sermons." Different initiatives by states and civil society on national level exist (Agency for Digital Government, 2019; Amnesty International, 2018; Association for Computing Machinery, 2017; Australian Human Rights Commission, 2019; Beijing Academy of Artificial Intelligence [BAAI], 2019; Commission Nationale de l'Informatique et des Libertés [CNIL] et al., 2018; Fairness, Accountability, and Transparency in Machine Learning [FAT/ML], 2016; Future of Life Institute, 2017; House of Commons of the United Kingdom – Science and Technology Committee, 2018; Japanese Society for Artificial Intelligence [JSAI], 2017; Montréal Declaration, 2018; Partnership on AI, 2024; The Public Voice Coalition, 2018; Swiss Federal Council, 2020; Villani, 2018).

On a regional level, states and civil society have contributed to this debate (European Group on Ethics in Science and New Technologies, 2009, 2014, 2015, 2018a, 2018b; High-Level Expert Group on Artificial Intelligence HLEG AI of the European Commission, 2019; Council of Europe, 2018a, 2018b, 2018c, 2019a, 2019b; European Union, 2018; *The Charter of Digital Rights* by European Digital Rights (EDRi)-network, 2014).

On an international level, contributions by state and non-state actors enter the spotlight (CNIL et al., 2018; Partnership on AI, 2024; The Public Voice Coalition, 2018; United Nations [UN] Educational, Scientific and Cultural Organization, World Commission on the Ethics of Scientific Knowledge and Technology [UNESCO & COMEST], 2017; Information Technology Industry Council [ITIC], 2017; Dutton, 1990; Price, 2018; D 64 Zentrum für Digitalen Fortschritt, 2018; United Nations Children's Fund [UNICEF], 2017; Association for Computing Machinery's Committee on Professional Ethics, 2018; WeGovNow, 2020; The Future of Privacy Forum, 2018; UNI Global Union, 2017; Institute of Electrical and Electronic Engineers [IEEE] Standards Association, 2024; Internet Governance Forum [IGF], 2014, 2019; ISO, 2020; the "Bletchley Declaration 2023" (AI Safety Summit, 2023); the "AI4People Summit Declaration"; the "G7 International Guiding Principles on AI"; the "AI Code of Conduct"; the "Recommendation on the Ethics of Artificial Intelligence" by UNESCO; the OECD Principles on Artificial Intelligence (OECD, 2019); the G20-Principles of AI; "Declaration of Principles" by the World Summit on the Information Society in Geneva 2003 (World Summit on the Information Society [WSIS], 2003). Contributions from professional ethics—e.g., the code of the Institute of Electrical and Electronic Engineers (IEEE Standards Association, 2024), the code of the National Society of Professional Engineers (NSPE, 2019), or the codes of the American Society of Mechanical Engineers (ASME, n.d.) emerge together with other proposals. The main challenge with these declarations, recommendations, principles, and guidelines—with "the sermons"—is the step from theory to practice. They remain at maximum "soft law" "which is a tool often used to either avoid or anticipate formal legislation" (Nevejans, 2016, p. 26).

Beyond that, the European Parliament and Council reached political agreement on the European Union's Artificial Intelligence Act ("EU AI Act") (European Parliament, 2023). The EU AI Act aims to represent a comprehensive legal framework for the regulation of AI systems across the EU, ensuring the safety of and respect for fundamental rights by DS as well as encouraging investments and innovation in the field of DS.

In addition, the Council of Europe Framework Convention on Artificial Intelligence and Human Rights, Democracy and the Rule of Law was adopted on May 17, 2024, by the Committee of Ministers of the Council of Europe (Council of Europe, 2024).

Other legal initiatives—forming together with the EU AI Act the second generation of governance initiatives: "the locals"—are also pursued, among others, in China and in the USA on the federal level, and several governments on the state level released new regulations, leading to the categorization of these activities as "American Market-Driven Regulatory Model," "Chinese State-Driven Regulatory Model," and "European Rights-Driven Regulatory Model" (Bradford, 2023, pp. 35–145).

Finally, the debate about global governance in DS also recognizes the third generation of governance initiatives: "the international players"—consisting of the proposals of two main models: the model of the Intergovernmental Panel on Climate Change (IPCC) (Kaspersen & Wallach, 2023) and the model of the International Civil Aviation Organization (ICAO) (Baker and McKenzie, 2023; World Economic Forum [WEF], 2022).

The model of the Intergovernmental Panel on Climate Change (IPCC) consists of a panel of experts. The IPCC was established in 1988 by the UN, with member countries from around the world. It provides governments with scientific information they can use to develop climate policies. The panel of experts in the domain of AI would provide policymakers and governments with information, scenarios, and models for their decision-making.

The model of the International Civil Aviation Organization (ICAO) (Baker and McKenzie, 2023) consists of a binding global framework and its implementation. The ICAO, as a UN agency, has its basis in the Convention on International Civil Aviation. The ICAO is the global forum of states for international civil aviation. It develops policies and standards, provides compliance audits, studies, analyses, and assistance to states and stakeholders, and contributes to the global alignment of air regulations.

While "the sermons" lack implementation during the working week, and "the locals" possess a national or regional focus, while DS represents a global phenomenon, "the international players" locate the governance of DS on an international level and approach it adequately in an institutional manner in order to create a positive impact in reality and on the ground.

At the same time, it needs to be considered that the IPCC model will not reach the desired effect on the ground because it possesses neither legal authority nor legal enforcement tools—it lacks "teeth." This weakness of the IPCC model becomes clear listening to the UN High Commissioner for Human Rights, Volker Türk:

> Victims and experts [...] have raised the alarm bell for quite some time, but policy makers and developers of AI have not acted enough – or fast enough – on those concerns. We need urgent action by governments and by companies. And at the international level, the United Nations can play a central role in convening key stakeholders and advising on progress. There is absolutely no time to waste. The world waited too long on climate change. We cannot afford to repeat that same mistake. (Turk, 2023)

Regarding the ICAO model, doubts arise about whether one can compare a self-contained industry like aviation to a cross-cutting technology like DS. The latter causes manifold and multifaceted legal and ethical issues (Baker and McKenzie, 2023).

The existing global governance initiatives and the proposals of two main models need to face the reality of global threats of digital transformation and DS.

9.3 Ten Global Threats of Digital Transformation and DS

From an ethical perspective, the *first global threat* of digital transformation and DS is not doing anything about the ethical risks of DS: Products with DS that violate human dignity and human rights or destroy the planet can be put on the market without legal consequences. For example, if an app is sold online that sexualizes images of children, nothing happens to the company offering this app except that it makes a lot of money (Heikkilae, 2022; Lensa, 2024; Snow, 2022). For several decades now, humanity has allowed business models that have human rights violations at their core to go unchallenged. Do the existing global governance initiatives and the proposals of two main models represent measures strong enough to counter these global dangers? It does not seem to be the case when one considers the following three test-questions:

1. Can human rights-violating state actions or business practices be concretely stopped because of these initiatives or proposals? No.
2. Do multinational technology companies need to change anything regarding their human rights-violating business practices because of these initiatives or proposals? No.
3. Can states or multinational technology companies be held accountable for their human violations based on these initiatives or proposals? No.

It becomes obvious that much more needs to be done. The threat of underestimating the impact of the massive reduction of paid professional tasks for humans and of not pushing forward SERT is another concrete and dramatic example of this global danger.

A *second global threat* posed by digital transformation and DS, as already mentioned above, is the growing inequality and poverty around the world due to a widening "digital divide" (Kirchschlaeger, 2021). Expected explosive increases in computing power (e.g., through quantum computing) will exacerbate the already existing "digital divide"—while in some regions of the world there is no access to the Internet (ITU, 2024), other places already benefit from large supercomputing centers (Swiss AI Association, n.d.).

A *third global threat* is the negative impact of digital transformation and DS on the climate and the environment, which threatens to destroy the planet (Kirchschlaeger, 2021). The massive demand for both raw materials for the production of DS and for energy in the production and use of DS is still disproportionate to the contribution of DS to combating climate destruction (AlgorithmWatch, 2025).

A *fourth global threat* is the constant violation of the human right to privacy and data protection (Zuboff, 2019). Whenever possible, people's data is stolen and sold to the highest bidder. The ongoing disregard for privacy and data protection is a massive attack on the freedom of all people.

A *fifth global threat* is that with digital transformation and DS, as elaborated above, thanks to these huge amounts of data about people, DS "know" people better than people know themselves. This opens the door to economic and political manipulation. With DS, you can know—to use a metaphor—exactly which piano keys to press to make the music play, in other words, to get people to buy or choose or vote the way the hands controlling DS want them to (Nosthoff & Maschewski, 2019).

A *sixth global danger* of digital transformation and DS is the potential for disinformation. Due to global networking online, millions of people can be targeted with false information—so-called "fake news"—at the click of a mouse. Language and image generation systems (such as ChatGPT, Dall-E), for which the criterion of truthfulness, the human rights to data protection and to privacy, and copyright play no role and which, figuratively speaking, are nothing more than—as introduced above in Chapter 1 "Introduction"—cows chewing the cud (Kirchschlaeger, 2024f), open up new horizons in this respect, as artificially created audio and video material with a high degree of perfection can be circulated globally as supposed realities. For example, words can be put into the mouths of politicians that they never actually said. Unfortunately, despite official clarifications from the politicians concerned, some of the untruths remain in people's minds—with "fake news" in images even more than in words (Kirchschlaeger, 2021).

Linked to this, a *seventh global threat* of digital transformation and DS is the possibility of targeted destabilization of societies through hate speech and incitement to violence spread on the Internet—especially on social media—where masses can be mobilized at the click of a mouse (Cadwalladr, 2024).

An *eighth global threat* of digital transformation and DS is the risk to the mental and physical health of people (Mahari & Pataranutaporn, 2024), especially children and adolescents, due to the effects of DS, including social media which actually are "anti-social media" (Kirchschlaeger, 2024c). Among other things, the WHO is considering recognizing social media use as an addiction (Montag, 2023) because social media has the following traits: "The internet is a giant hypodermic, and the content, including social media like Meta, are the psychoactive drugs" (Richtel, 2023).

A *ninth global threat* of digital transformation and DS is cyberattacks (Kirchschlaeger, 2021). Growing digitalization and an increase in the use of DS in people's business and private lives, as well as in the systemic infrastructure of societies, mean an increase in vulnerability to cyberattacks. Cybercrime and the fight against it have a massive economic impact on national economies.

The military applications of digital transformation and DS represent a *tenth global threat* (Kirchschlaeger, 2024b). Automated weapon systems are unable to distinguish between combatants and non-combatants in armed conflicts and to apply the principle of proportionality (Asaro, 2008), which leads to more misconduct and crime (Sharkey, 2011). Another ethical argument against automated weapons is that they would lead to more wars (Kahn, 2017) because, among other things, a lower number of victims would be expected (Wallach, 2015) and thus a lower political price would have to be paid (Wagner, 2014).

Facing these *ten global threats*, the limitations of the so-far three generations of governance initiatives become obvious when one approaches this risky reality with the three test-questions mentioned above for the already existing "sermons," "locals," and "international actors." The already existing "sermons," "locals," and "international actors" receive negative responses in all three cases. Therefore, it is necessary to think about ways to combine the strengths of "the sermons," "the locals," and "the international actors" and to avoid their weaknesses in order to address these global threats in an impactful and global way while still benefiting from the ethical opportunities of DS. Allowing humans and the planet to flourish sustainably and guaranteeing globally that human dignity is respected not only offline but also online and in the digital sphere as well as in the domain of DS, the two concrete measures described below are proposed.

9.4 Human Rights-Based Data-Based Systems (HRBDS)

In order to promote human rights—including especially the rights of the child (Kirchschlaeger, 2025)—as the guiding ethical principle as well as the legal basis for SERT, a regulatory framework for data-based systems (DS) based on human rights should be achieved. Human rights-based data-based systems (HRBDS) (Kirchschlaeger, 2021, 2024a, 2024d, 2024e) are meant to ensure that human rights serve as the basis of DS. In other words, HRBDS seek to ensure that human rights are respected, protected, implemented, and realized within the entire life cycle of DS and the complete value chain process of DS (in the design, the development, the production, the distribution, the use, or the non-use of DS because of human rights concerns). HRBDS strives for protecting the powerless from the powerful.

As a minimum ethical frame of reference, human rights could provide the necessary normative orientation in that they only represent a minimum standard that guarantees all humans physical survival and a life with human dignity—a life as a human. As an ethical minimum standard, human rights do not ethically overtax technological progress and do not stifle it, as would be the case with a high ethos. Another argument in favor of human rights is the ethical justifiability of human rights and their universality with the principle of vulnerability (Kirchschlaeger, 2013c), which underpins their global acceptance. The latter proves to be unrivalled in comparison to other value systems and catalogs of norms (Kirchschlaeger, 2016), as human rights are not anchored in a specific culture, tradition, religion, or worldview but serve as a basis for the lived diversity of cultures, traditions, religions, or worldviews by protecting, among other things, the freedom of belief, conscience, and religion of all people (Kirchschlaeger, 2013b, 2020).

In addition, human rights encourage and promote innovation and technological progress by protecting freedom of thought, freedom of expression, and free access to information (Kirchschlaeger, 2013a) which are all essential and constitutive for optimal research-and-innovation processes. In other words, it will be impossible to be innovative if one is not allowed to express one's opinion freely, if one does not enjoy access to all information available, and if one is not allowed to question authorities.

Finally, they protect and promote the diversity that is constitutive for innovation processes by guaranteeing the self-determination of every person, which proves to be a source of diversity (Kirchschlaeger, 2016, 2020).

Amid the ten global dangers of digital transformation and DS explored above, more and stricter commitment to a human rights-based legal framework is necessary and regulation that is precise, goal-oriented, and strictly enforced. HRBDS would serve this necessity. In this way, regulation may also be advantageous economically and fosters innovation and entrepreneurship guaranteeing legal security as well as planning reliability for ventures and companies. For example, the American regulation of air traffic and the aviation industry allowed an entire industry to flourish economically thanks to its high degree of precision, its clear orientation, and its uncompromising enforcement (Selz, 2019). Regulation is nothing new for technological progress.

> This is also how our societies proceeded in the 20th century. When industrialization began, most countries had no health and safety regulations, no employee rights. These rights had to be fought for before industrial societies were finally shaped in such a way that they could also be democratic societies. In the same way, we now have to build new institutions – and think about what rights we need to codify so that people's autonomy is protected, so that they can continue to enjoy freedom and privacy. (Zuboff, as cited in Weisbrod, 2025, p. 43)

9.5 International Data-Based Systems Agency (IDA) at the UN

In order to promote human rights-based data-based systems (HRBDS) allowing humans and the planet to flourish sustainably and guaranteeing that human rights are respected globally not only *offline* but also *online*—and in the digital sphere and the domain of DS—an International Data-Based Systems Agency (IDA) at the UN (Kirchschlaeger, 2021, 2024a, 2024d, 2024e) urgently needs to be established as a global platform for technical cooperation in the field of DS, fostering human rights, safety, security, a rules-based international order, sustainability, and peaceful uses of DS with a multilateral approach and by collaborating and joining forces with other UN agencies and already existing events and formats, e.g., the World Summit on the Information Society WSIS, the International Governance Forum IGF, the AI for Good Global Summit, as well as a global supervisory and monitoring institution and regulatory authority in the area of DS responsible for access to market approval. Mustafa Suleyman (CEO of Microsoft AI, Co-Founder and former Head of applied AI at DeepMind, Co-Founder of Inflection AI) says:

> We need our generation's equivalent of the nuclear treaty to shape a common worldwide approach – in this case not curbing proliferation altogether but setting limits and building frameworks for management and mitigation that, like the wave, cross borders. This would put clear limits on what work is undertaken, mediate among national licensing efforts, and create a framework for reviewing both. Where there is a clear scope for a new body or bodies is with technical concerns. A dedicated regulator that navigates contentious geopolitics (as much as possible), avoids overreach, and performs pragmatic monitoring function on broadly objective criteria is urgently needed. Think of something like the International Atomic Energy Agency. (Suleyman & Bhaskar, 2023, p. 266)

The aim of IDA at the UN is to ensure and to promote the development and deployment of HRBDS as a regulatory framework guaranteeing the use of the ethical positive potential of DS for the benefit of all humans and the planet as well as the handling of its ethical negative potential, including the destruction of humankind and the planet. Serving this aim is the establishment of robust governance mechanisms.

Given the areas of convergence between DS and nuclear technologies, the International Atomic Energy Agency (IAEA) model would seem the most appropriate one for responsible global AI governance as it represents an UN-agency with "teeth." IDA needs to be built following the model of the International Atomic Energy Agency (IAEA) as an "institution with teeth" because, thanks to its legal powers, functions, enforcement mechanisms, and instruments, the IAEA was able to foster innovation and ethical opportunities while at the same time protecting humanity and the planet from the existential risks in the domain of nuclear technologies, which also embrace the same dual nature as DS, covering both ethical upsides and downsides. Lessons from the IAEA experience underscore the importance of prioritizing human rights, transparency, and accountability in DS governance frameworks. By combining a human rights-based approach to DS development and deployment with establishing IDA as a central regulatory authority, the international community can proactively address existential AI risks while harnessing the transformative potential of DS to advance human rights and sustainability. Leveraging the lessons learned from nuclear technologies and the establishment of the IAEA, the establishment of IDA presents a viable pathway towards effective global governance of existential AI risks, ensuring the responsible and ethical development of DS for the betterment of humanity and the planet.

Based on human rights and on the United Nations' Sustainable Development Goals, as well as in critical dialogue with already existing soft-law instruments and declarations (Montreal Declaration, 2018; CNIL et al., 2018; Future of Life Institute, 2017; The Public Voice Coalition, 2018; Amnesty International, 2018; IEEE Standards Association, 2024; Partnership on AI, 2024), the following 30 principles should provide ethical guidance to IDA (Kirchschlaeger, 2021, 2024a). The *30 IDA principles* consist of more principles in order to address the most recent ethical challenges as well as lacunae, which one could argue for in the so-far existing soft-law instruments and declarations. For example, it is not enough to call for transparency because if someone acts illegitimately, the unethical quality of this action is not transformed into something legitimate just due to transparency. Similarly, it is not enough that DS are explainable. DS must also be intelligible in order to enable humans to benefit from the ethical opportunities and to master the ethical risks of DS.

The IDA should serve the realization of the following *30 IDA principles*:

First principle: Data-based systems must respect, protect, implement, and serve the realization of human rights.

This principle ensures that the respect, protection, implementation, and realization of human rights are guaranteed not only "offline" but also "online," in the digital sphere, and in the domain of DS. Especially this first principle—together with the following 29 IDA principles—addresses the above-outlined ten global threats because they ensure that human rights are respected, protected, implemented, and realized throughout the entire life cycle of DS.

Second principle: Data-based systems must serve the realization of the United Nations' Sustainable Development Goals.

Beyond their contribution to the realization of human rights, this principle guarantees that DS foster the realization of the UN SDGs.

Third principle: Data-based systems must be transparent.

This principle ensures the transparency of DS, which means that the inner workings of DS are accessible. Humans should not blindly trust DS because they are neither objective, fair, nor neutral, but their algorithms and the data they are relying on are neither objective, fair, nor neutral, but biased.

Fourth principle: Data-based systems must be traceable.

Combined with the third principle, this principle guarantees that the steps of the inner-working process of DS are not only accessible, but their traces are also identifiable, allowing a precise and specific account of the single steps within a DS.

Both—the third principle and the fourth principle—build the foundation of the possibility for humans to bear their responsibility for DS. This is necessary because DS do not possess moral capability due to a lack of the principles of vulnerability, conscience, freedom, and autonomy. Thus, they cannot be responsible for their decisions and actions.

Fifth principle: Data-based systems must be explainable.

This principle ensures the possibility of an explanation of the inner-working process of DS which allows humans to learn about it.

Sixth principle: Data-based systems must be intelligible.

Combined with the fifth principle, this principle guarantees that humans not only can know about the specific steps of the inner-working process of DS, but that humans are also able to understand and act upon it.

The fifth principle and the sixth principle represent further pillars of the fundament of the possibility for humans to bear their responsibility for DS because humans cannot perform their responsibility for DS if explanations of the inner-working process of DS as well as intellectual access to the inner-working process of DS would not be provided to them.

Seventh principle: Data-based systems must be auditable.

Combined with the third principle and the fourth principle, this principle—based on transparency and traceability of DS—opens the horizon for auditing DS, allowing for the identification of responsibilities and liabilities. This is the only way to empower humans to restore justice and to enable humans to address possible accidents or mistakes by DS in an adequate manner, which increases, among others, the probability that the same accidents and mistakes will not happen again.

Eighth principle: Causes and effects, or causality and correlation, must be identifiable in data-based systems.

This principle ensures that confusions of causes and effects as well as of causality and correlation in DS—both scenarios potentially deleting essential differentiations in cognitive perception and both scenarios potentially resulting in blurriness of responsibility and leading to conceptual and epistemological blindness as well as terrible consequences—can be avoided by identifying them precisely in DS.

Ninth principle: Data-based systems must be predictable.

Combined with the eighth principle, this principle empowers humans in their decision-making to trust the assistance, support, and work of DS or not. Only this way humans can live up to their responsibility for DS because they understand what will happen if they use DS as well as what will happen if they do not deploy DS.

This leads to the specific ethical consequence that self-learning DS must be designed and built in a way that full predictability is given. This is not surprising because the usual necessity of humans is that they use machines only if they know what they can expect is going to happen. Or would one, e.g., board an aircraft if one did not know that the aircraft was normally able to fly?

Tenth principle: Data-based systems must be decidable.

This principle guarantees that DS can take decisions and that the opinion-forming and decision-making processes resulting in decisions can be accessed and followed by humans.

Eleventh principle: Data-based systems must be non-manipulating and respect the autonomy of every human.

This principle explicitly addresses the ethical danger of manipulation of humans by DS as well as disrespect of the autonomy of humans by DS. DS can spread disinformation, "fake news", and "deep fakes" undermining the freedom and autonomy of humans as well as the credibility of human opinion-forming and decision-making processes. Based on the enormous amount of data about humans, DS "know" humans better than humans know themselves. Humans are therefore open victims of manipulation. One can get people using DS to vote the way one wants them to vote. Metaphorically speaking, DS "know" exactly which keys of the piano they must play so that the music resounds.

Twelfth principle: Data-based systems must be able to adapt to humans.

This principle guarantees that DS are human-centered, and that DS serve humans, not vice versa. At the same time, it makes sure that DS are inclusive.

Thirteenth principle: Data-based systems and their performance (efficiency and effectiveness) must be controlled, monitored, measured, and evaluated on a regular basis, and their assessment must be published each time such that it is accessible to the broader public.

This principle guarantees that the ethical upsides and ethical downsides of DS are adequately addressed in relation to their essential and fundamental nature for humans and that public and transparent participatory and democratic monitoring, oversight, and handling are in place out of respect for the human dignity and freedom of all humans.

Fourteenth principle: Data-based systems must include an "emergency button" (metaphorically) for humans and an "ethics-black-box" enabling an ethical analysis.

This principle ensures, with its first part, that humans remain in charge due to their exclusive responsibility for DS as introduced above and that humanity and the planet are prepared for a worst-case scenario. With its second part, it empowers humans to review decisions and actions by DS and evaluate them from an ethical point of view.

Fifteenth principle: Data-based systems must be approved by international and national regulatory authorities—similar to food and drug regulatory agencies protecting public health by ensuring the safety, efficacy, security, and sustainability of data-based systems, regulating the manufacturing, marketing, and distribution of data-based systems, helping to further innovations that make data-based systems more effective, safer, and more affordable, and empowering the public by providing the accurate, independent, and science-based information they need to accept and use data-based systems.

This principle guarantees that only human rights-respecting and sustainable DS can be put on the market—an aspect that is naturally guaranteed in other industries (e.g., in the pharmaceutical industry, of course a medical drug can only be brought to market after a careful approval process that excludes harm to nature and people)—and that human rights and sustainability continue to be respected, implemented, protected, and realized in the digitalized value chains as well as in the entire life cycle of DS.

Sixteenth principle: Research and development projects in the area of data-based systems must be approved by international and national regulatory authorities.

This principle ensures that research and development in such a high-risk area as DS happens within a well-defined legal and ethical framework based on the freedom and independence of research. For example, in the area of nuclear technologies—knowing a similar dual nature and the potential for an existential impact as DS—research and development are strictly regulated in order to guarantee the flourishing of humanity and the planet, and this regulation is implemented and enforced resolutely by the IAEA.

Seventeenth principle: The conduct of research and development must respect these IDA principles.

This principle makes sure that present and future DS honor these IDA principles.

Eighteenth principle: Lethal automated weapons and lethal automated weapon systems are forbidden.

This principle guarantees the prohibition of lethal automated weapons and lethal automated weapon systems because automated weapons face fundamental criticism from an ethical standpoint (Kirchschlaeger, 2024b) in that they are not able, according to the laws, to distinguish in armed conflicts between combatants and non-combatants and to apply the principle of proportionality, resulting in more wrong actions and crimes. Beyond that, the "strategic robot problem" (Roff, 2014) and its ethical implications underpin the idea that automated weapons are ethically problematic. The "strategic robot problem" lies in the undermining of command-and-control structures by creating automated weapons to serve as combatants and commanders at the same time. From an ethical viewpoint, the following factors would also speak against automated weapons: the idea that they would lead to more wars (Kahn, 2017) because of the reduction of "ability to make credible threats and assurances in a crisis" (Leys, 2018), the growing distance between human actions and their consequences, (Pappenberger, 2013; Singer, 2009) less human involvement for the actors deploying them (Grut, 2013), a lower number of victims would be expected (Leys, 2018; Wallach, 2015), and a lower political price would need to be paid (Wagner, 2014). The latter becomes obvious for reasons including but not limited to the replacement of the fundamental reciprocity of combat, namely, possessing as a soldier the power to kill while running the ongoing risk of being killed (Broeckling, 2015; Hood, 2015; Kaufmann, 2010). As a consequence of the latter, legislative oversight would not be respected, and this would work against the system of checks and balances.

Nineteenth principle: Data-based systems for human rights-violating surveillance are halted and forbidden.

This principle ensures that these already existing and already deployed human rights-violating DS are stopped and prohibited.

Twentieth principle: Data-based systems for social scoring of humans by the state or by non-state actors are halted and forbidden.

This principle ensures that these already existing and already deployed human rights-violating DS are stopped and prohibited.

Twenty-first principle: Data-based systems manipulating and undermining democracy are halted and forbidden.

This principle ensures that these already existing and already deployed anti-democratic DS are discontinued and prohibited.

Twenty-second principle: Data-based systems supporting or reinforcing totalitarian systems and dictatorships are halted and forbidden.

This principle ensures that these already existing and already deployed anti-democratic DS are terminated and prohibited.

Twenty-third principle: Data-based systems blazing the trail for "super-data-based systems," or "singularity" are forbidden.

This principle guarantees that technological progress, which would lead to "super-DS" or "singularity"—both endangering humanity and the planet—is still stopped at a moment when humans are still in control and prohibited.

Twenty-fourth principle: "Super-data-based systems" or "singularity" are forbidden.

This principle ensures the flourishing of all humans and the planet by prohibiting "super-DS" or "singularity" because they represent existential threats to humanity and the ecosystem.

Twenty-fifth principle: These principles so far must be included in the parameter settings for the creation, design, programming, development, production, training, and use of data-based systems.

This principle guarantees that the IDA principles are included in the creation, design, programming, development, production, training, and use of data-based systems.

Twenty-sixth principle: Designers, software engineers, manufacturers, producers, operators, providers, and users of data-based systems, as well as infrastructure providers and data analytics companies and their employees, must have adequate knowledge, skills, and competencies, including a basic applied ethics expertise.

This principle ensures that the professionals involved in the creation, design, programming, development, production, training, and use of data-based systems—because of their high relevance to the background of the IDA principles—possess adequate knowledge, skills, and competencies, also in applied ethics.

Twenty-seventh principle: Designers, software engineers, manufacturers, producers, operators, providers, and users of data-based systems, as well as infrastructure providers and data analytics companies and their employees, must be accountable. They must be able to take legal and ethical responsibility.

This principle guarantees the legal and ethical empowerment of the professionals involved in the creation, design, programming, development, production, training, and use of data-based systems because of their crucial role in the realization of the IDA principles.

Twenty-eighth principle: The principle of indivisibility of all IDA principles must be respected.

This principle ensures that all IDA principles complement each other, go hand in hand, and that they are all optimally realized.

Twenty-ninth principle: Any supplement or modification to these principles must be undertaken only by humans.

This principle guarantees the respect of the exclusive responsibility of humans for DS.

Thirtieth principle: Any supplement or modification to these principles must undoubtedly serve the realization of human rights for all humans and the United Nations' Sustainable Development Goals.

This principle ensures that any necessary change to the IDA principles contributes to the realization of human rights of all humans and the United Nations' Sustainable Development Goals.

Compared to other models for global governance of DS, IDA promises to reach the precision, goal orientation, and strict enforcement not only necessary to guarantee the flourishing of humanity and the planet from an ethical standpoint but also to foster innovation and entrepreneurship from an economic point of view.

What makes the establishment of an IDA realistic is not only its essential and minimum normative framework, its practice-oriented and participatory governance-structure, which will be introduced below in more detail, as well as its striving for legitimacy combined with fostering innovation but also that in the past, humanity has shown that when the well-being of people and the planet is at stake, humanity can focus on what is technically feasible rather than blindly pursuing all that is technically possible.

Humanity did pursue nuclear technology, develop the atomic bomb, and even deploy it more than once. But to prevent yet worse events, humanity then massively restricted the research and development of nuclear technology despite overwhelming opposition by state and non-state actors. That nothing worse has happened is largely due to global regulation, concrete enforcement mechanisms, and the International Atomic Energy Agency (IAEA) of the UN (Kirchschlaeger, 2021).

In the case of chlorofluorocarbons (CFCs), humanity also decided under the Montreal Protocol of 1987 (UN Environment Programme, 1987) to ban substances that damage the ozone layer and to enforce the ban consistently. Here, the resistance was also huge, inter alia, due to special interests from the private sector. This regulation and its uncompromising enforcement led to the fact that the hole in the ozone layer is now slowly closing.

Beyond that, DS distinguish themselves from nuclear technology and from CFCs especially in three characteristics that increase the realizability of the establishment and the existential impact of IDA for humanity and the planet:

1. In order to function, DS must have power. This means that if a DS is violating human rights, threatening peace, or destroying the planet, it can be stopped by taking it off the power grid or by cutting off the power supply.

2. In order to function, DS must be connected because of its dependence on data flow. This means that if a DS is violating human rights, threatening peace, or destroying the planet, it can be stopped by disconnecting it.
3. While operating, every DS leaves data traces, allowing identification and accountability.

Finally, it also builds an advantage over attempts to ensure that all humans benefit from a previous technology-based innovation and to master the ethical dangers of a previous technology-based innovation (like in the case of nuclear technologies with the establishment of the International Atomic Energy Agency or in the case of chlorofluorocarbons (CFCs) with the Montreal Protocol of 1987 (UN Environment Programme, 1987)), IDA could also rely on DS-based solutions to implement HRBDS.

As introduced above, the line of argumentation showing the necessity of the establishment of IDA at the UN includes an analogy with the International Atomic Energy Agency (IAEA). These first preliminary thoughts about the basis of IDA (Kirchschlaeger, 2024a) developed in the following sections are inspired by the legal architecture of the IAEA (International Atomic Energy Agency, 2011, 2013; Stoiber et al., 2003; Rockwood, 2013; El Baradei et al., 1995; Sharma, 1995).

The legal basis for the establishment of IDA should be a UN resolution elaborating and adopting the text of the Statute of IDA, which constitutes the following elements:

a. Purpose: The purpose of IDA is—as defined above—to be a platform for technical cooperation in the field of digital transformation and DS, fostering human rights, safety, security, and peaceful uses of DS; and to act as a global supervisory and monitoring institution and regulatory authority in the area of DS responsible for access to market approval, partnering with and supporting on a global level the work of the national regulatory authorities in the area of digital transformation and DS. It should foster the safe, secure, and peaceful use of data-based systems, contributing to international peace and security, the respect and realization of human rights, and the United Nations' Sustainable Development Goals.
b. 30 IDA principles (please see above).
c. Legal Status of IDA (please see below).
d. Membership in the IDA (please see below).

e. Rights and Responsibilities of IDA (please see below).
f. Mechanisms, Measures, and Instruments of IDA (please see below).
g. Governance of IDA (please see below).

While IDA is not a state under international law, it is an entity with "international legal personality." States should recognize IDA as an entity that has some rights and privileges normally associated with a sovereign state. One of the primary attributes of an international organization such as IDA is its capacity to conclude international agreements with other "persons" having international legal personality under international law.

Beyond that, the legal status of IDA also embraces the relationship of IDA with the UN, including its regular reports to the UN General Assembly as well as, if necessary, to the UN Security Council and its institutional cooperation within its own decision-making processes.

These points ensure the ability of IDA to act and have a concrete impact, as well as its checks and balances.

State and non-state actors become members of the IDA by ratifying its statute.

Offering non-state actors membership in IDA is necessary due to the growing political importance of multi-stakeholder participation, the increasing economic power (and corresponding responsibility) of multinational technology companies in the field of DS, as well as the global and multidisciplinary nature of DS.

Vis-à-vis state and non-state actors (e.g., corporations), IDA is authorized to establish and administer safeguards to foster the realization of the 30 IDA principles, among others, and primarily,

- To guarantee that DS are developed, produced, and deployed with respect to human rights;
- To ensure that DS are promoting peace;
- To secure that DS are fostering the realization of the UN SDGs;
- To apply safeguards, at the request of the parties, to any bilateral or multilateral arrangement;
- To apply safeguards to any of the DS activities of a state or non-state actor at that state actor's or non-state actor's request.

IDA enjoys rights and responsibilities, including, among others:

- The right to permit or prohibit as a global regulatory authority research and development-projects of DS in order to protect all humans and the planet (in analogy with such legally defined approval processes in the pharmaceutical industry protecting humans and the planet);
- The right to examine, as a global supervisory and monitoring institution, the design, research, and development of DS;
- The right to decide as a global regulatory authority on the approval of access to the market of a DS (in analogy with such legally defined approval processes in the pharmaceutical industry protecting humans and the planet);
- The right to examine the production of DS as a global supervisory and monitoring institution;
- The right to examine the deployment of DS, including requesting operating records to assist in ensuring accountability for and control of DS as a global supervisory and monitoring institution;
- The right to require the submission of reports on the design, production, and/or deployment of DS regarding human rights, peace, and sustainability as a global supervisory and monitoring institutions;
- The right to send as global supervisory and monitoring institution into the state actor or to the non-state actor members of the "enforcement committees" and/or inspectors, designated by IDA after consultation with the state actor or state actors, respectively, non-state actor or non-state actors concerned, who shall have access at all times to all places and data, and to any person who, by reason of his or her occupation, deals with DS;
- In the event of non-compliance and failure by the state or non-state actor concerned to take as global supervisory and monitoring institution requested corrective steps within a reasonable time, the right to curtail or suspend assistance and call for the disconnection and deactivation of DS combined with the enforcement of corresponding fines (in percentages of the annual budget of the state, respectively, the profit before tax of a non-state actor) and with the execution of the suspension of the state or non-state actor from the exercise of the privileges and rights of IDA membership;
- The responsibility to serve as a platform for technical cooperation in the field of digital transformation and DS—of course pursing a multistakeholder-approach and collaborating and joining forces with

already existing events and formats (e.g., the World Summit on the Information Society (WSIS), the International Governance Forum (IGF), and the AI for Good Global Summit)—fostering human rights, safety, security, and peaceful uses of DS.

These rights and responsibilities of IDA serve the goal that IDA realizes a concrete and sustainable ethically positive impact in the domain entrusted to IDA.

Legally binding instruments as part of international law should be developed under the auspices of IDA, by UN member states, fostering the enjoyment of ethical opportunities by all humans as well as mastering or avoiding ethical risks.

The process of concluding a legally binding instrument is as follows:

- Either the usual process of UN treaties can be started by either a state actor, a civil society actor, or IDA
- Or—in the case of urgency or emergency regarding the respect of human rights, peace, and sustainability—by IDA.

Every legally binding instrument will have an enforcement committee consisting of independent experts examining the implementation by the state and non-state actors based on annual reports by the state and non-state actors. The work of the enforcement committee is also informed by visits as well as inspectors equipped with enforcement-measures, more specifically:

- Addressing the concrete opportunities and risks as well as the measure calling upon the state or non-state actor to remedy non-compliance;
- Reporting non-compliance to the UN Member states and to the UN Security Council and the UN General Assembly;
- Defining countermeasures against the state or non-state actor, including fines (in percentages of the annual budget of a state, respectively, the profit before tax of a non-state actor) and the suspension of the state or non-state actor from the exercise of the privileges and rights of IDA membership.

Non-binding instruments (e.g., IDA recommendations, IDA guidelines, and IDA code of conducts) should be developed under the auspices of IDA for the promotion of human rights, peace, and sustainability even beyond the letter of the law, as well as addressing in the first phase new technology-based ethical opportunities and risks potentially inspiring new legally binding instruments. Every non-binding instrument will be accompanied by an advisory group consisting of independent experts consulting state and non-state actors, receiving and reviewing the reports of state and non-state actors in line with the conceptualization of IDA as a learning organization, which is of utmost importance in the area of rapidly evolving technology-based innovation.

Already existing regional and national legally binding instruments should be promoted by IDA as good practices for other regions or states, including general comments evaluating them based on the purpose of IDA and the 30 IDA principles.

Already existing regional and national non-binding instruments should be promoted by IDA as good practices for other regions or states, including general comments evaluating them based on the purpose of IDA and 30 IDA principles.

These instruments, enforcement mechanisms, and enforcement measures of IDA serve the aim of creating, in an inclusive and participatory way, the aspired concrete effects in the field of DS.

IDA governance shall include

a. Director General (an individual endowed with executive powers), elected by the UN General Assembly;
b. Secretariat, elected by the Director General and confirmed by the Board (implementing body);
c. Board (strategic body), elected by the UN General Assembly;
d. Tripartite Council consisting of three representatives of states, of the private sector, and of civil society (advisory body to the Director General and the Board), confirmed by the UN General Assembly;
e. UN Council consisting of the UN Secretariat General and all UN Agencies—especially UNESCO formulating the first global ethical standards for so-called Artificial Intelligence (collaborative body for the Director General and the Board);
f. Enforcement committees, elected by the UN General Assembly (executive bodies);
g. Advisory groups, elected by the UN General Assembly (consulting bodies for state and non-state actors);

h. Inspectors team, appointed by the Director General and the relevant enforcement committee (monitoring body).

This governance structure of IDA ensures its ability to decide, act, and deliver concrete results and impact on the ground, its good governance, its embeddedness in the UN and the international community, as well as its interconnectedness with other international institutions and organizations in this field, its multi-stakeholder participation, its inclusive approach integrating and benefiting from all the work and efforts already pursued and continuing to be pursued in this field, its excellent expertise, know-how, and experience, as well as its ability to implement and enforce according to its mandate.

HRBDS as innovative technology-based solutions should support IDA in fulfilling its responsibilities as far as possible (ethical problems that could occur during the deployment of DS, e.g., contextual bias, should be foreseen as much as possible), e.g., by identifying human rights-violating DS such as racist or sexist apps. (In order to illustrate this point, if DS are able to master the complexity of identifying malignant tumor cells on a screen, it must also be possible to use DS to identify available racist or sexist apps.) This means concretely, e.g., that a state or non-state actor needs to submit digitally an application for its research and development project in the field of DS before starting it, and the assessment and evaluation before the 30 IDA principles and the legal requirements of this project can be performed by DS informing a final decision by humans.

An international fund should be established by state and non-state actors providing impact-investments in ventures striving to bring HRBDS on the market, supporting IDA in fulfilling its responsibilities.

There is a growing international and interdisciplinary network of experts calling for the establishment of HRBDS and IDA (www.idaonline.ch). "The Elders," an independent group of world leaders founded by Nelson Mandela that includes former UN Secretary General Ban Ki-moon, Ellen Johnson Sirleaf (former President of Liberia and Africa's first elected female head of state as well as Nobel Peace Laureate), and Ireland's first female President Mary Robinson, have endorsed the recommendations for human rights-based DS and for the establishment of a global agency like IDA to monitor them. It has called upon the UN to take appropriate action. In their statement of May 31, 2023, the Elders declared that:

> A new global architecture is needed to manage these powerful technologies within robust safety protocols, drawing on the model of the Nuclear Non-Proliferation Treaty and the International Atomic Energy Agency. These guardrails must ensure AI is used in ways consistent with international law and human rights treaties. AI's benefits must also be shared with poorer countries. No existing international agency has the mandate and expertise to do all this. The Elders now encourage a country or group of countries to request as a matter of priority, via the UN General Assembly, that the International Law Commission draft an international treaty establishing a new international AI safety agency. (The Elders, 2023)

The idea of a human rights-based and legally binding regulatory framework as well as the establishment of an institution enforcing the global regulation enjoys the support of Pope Francis (Francesco, 2023).

His Holiness the Dalai Lama supports human rights-based data based systems (HRBDS) and the establishment of an International Data-Based Systems Agency (IDA) at the UN because HRBDS and the establishment of IDA at the UN foster the realization of human rights, peace, and sustainability in the sphere of data-based systems (DS) (Dalai Lama, 2025).

UN Secretary General António Guterres also supports the creation of an international AI watchdog body like the International Atomic Energy Agency (IAEA): "I would be favorable to the idea that we could have an artificial intelligence agency [...] inspired by what the international agency of atomic energy is today" (Guterres, as cited in Nichols, 2023). He has called for a new UN body like an "International Data-Based Systems Agency (IDA)" to tackle threats posed by artificial intelligence in the UN Security Council on July 18, 2023 (Guterres, 2023).

UN High Commissioner for Human Rights Volker Türk has demanded at the UN Human Rights Council "urgent action" and proposed human-rights-based HRBDS and a coordinated global response towards an institutional solution like the creation of an "International Data-Based Systems Agency (IDA)" in his statement about AI and human rights on July 12, 2023 (Türk, 2023).

Also, some voices from multinational technology-companies—among others, Sam Altman (Founder of OpenAI) at the World Economic Forum WEF 2024 (Santelli, 2024), and Mustafa Suleyman (CEO of Microsoft AI, Co-Founder and former Head of applied AI at DeepMind, Co-Founder of Inflection AI) (Suleyman & Bhaskar, 2023)—have called for IDA.

First foundations have been laid for the future path in the right direction: The UN Human Rights Council unanimously adopted, on July 14, 2023, its latest resolution on "New and emerging digital technologies and human rights" (UN Human Rights Council, 2023) which addressed for the first time the promotion and protection of human rights with explicit reference to "AI." The Resolution emphasizes that new and emerging technologies with an impact on human rights "may lack adequate regulation," highlighted the "need for effective measures to prevent, mitigate and remedy adverse human rights impacts of such technologies" and stressed the need to respect, protect, and promote human rights "throughout the lifecycle of artificial intelligence systems." It called for frameworks for impact assessments related to human rights, for due diligence to assess, prevent, and mitigate adverse human rights impact, and to ensure effective remedies, human oversight, and accountability.

On March 21, 2024, the UN General Assembly unanimously adopted a resolution "Seizing the opportunities of safe, secure and trustworthy artificial intelligence systems for sustainable development" (UN General Assembly, 2024) on the promotion of "safe, secure and trustworthy" "artificial intelligence (AI)" systems that will also benefit sustainable development for all. It emphasizes: "The same rights that people have offline must also be protected online, including throughout the life cycle of artificial intelligence systems."

On September 22, 2024, the UN General Assembly adopted the "Pact for the Future" including a "Global Digital Compact" and a "Declaration on Future Generations" (UN, 2024). These are steps in the right direction and represent first significant foundations for a future where all humans can live a life with human dignity and the planet enjoys sustainability.

At the same time, it becomes clear if one poses the following test-questions mentioned above that there are only first small steps in the right direction and that much more needs to be done:

1. Can human rights-violating state actions or business practices be concretely stopped because of these steps? No.
2. Do multinational technology companies need to change anything regarding their human rights-violating business practices because of these steps? No.
3. Can states or multinational technology companies be held accountable for their human violations based on these steps? No.

9.6 It Is Urgent Time to Act

These negative responses, as well as the following realities, underline that it is urgent time to act:

In the year 2025, it is still possible, e.g., as mentioned above to put on the market an app that sexualizes children's images (Heikkilae, 2022; Lensa, 2024), and the only thing that happens to this company is that it makes a lot of money from it. We need to do something about this! Urgent action is needed.

In the year 2025, it is now possible with "social media" to destabilize within a few hours a peaceful, functioning, and wealthy country (e.g., the UK far-right riots 2024 (Cadwalladr, 2024)).

In the year 2025, it is now possible with "Generative AI" to destroy a politician with a "Hollywood-quality" deep fake (Verma & Zakrzewski, 2024).

In the year 2025, the energy consumption by DS is exploding (Ren & Wierman, 2024).

In the year 2025, the global "digital divide" is growing (UN and International Labour Organization, 2024).

DS have ethical implications: opportunities as well as risks. DS can be powerful for fostering human rights and sustainability, but can, at the same time, be dangerous in violating human rights and in its ecological negative impact. The late Stephen Hawking pointed out:

> Unless we learn how to prepare for, and avoid, the potential risks, AI could be the worst event in the history of our civilization. It brings dangers, like powerful autonomous weapons, or new ways for the few to oppress the many. It could bring great disruption to our economy. (Kharpal, 2017)

It is widely recognized that DS have profound, dynamic positive and negative effects on societies, ecosystems, human lives, and especially on young people, adolescents, and children of today and of the future as they affect human thinking and interaction, science and culture, education, information, and communication. In order to allow young people, adolescents, and children of today and of the future to flourish on a planet with a sustainable future, it is necessary to address the ethical questions in the field of DS. UN Secretary General António Guterres pointed out in the UN Security Council on July 18, 2023: "Generative AI has enormous potential for good and evil at scale. [...] Without action to address these risks, we are derelict in our responsibilities to present and future generations" (Guterres, 2023). Human rights-based data-based systems (HRBDS) and the establishment of an International Data-Based Systems Agency (IDA) at the UN would create a solid fundament and a supportive framework for SERT serving a flourishing future with DS for humanity and for the planet.

It's time to act before even more people experience violations of their human rights with or by DS.

It's time to act before the environment and the climate get even more destroyed due to the exploding ecological footprint of DS.

It's time to act before there isn't any possibility to act anymore.

It's time to act before DS act for, on, and against humans.

It's time to act before it's too late.

References

Agency for Digital Government. (2019, March). *Danish National Strategy for Artificial Intelligence*. Retrieved March 31, 2025, from https://en.digst.dk/strategy/the-danish-national-strategy-for-artificial-intelligence/

AI Safety Summit. (2023). *The Bletchley declaration by countries attending the AI Safety Summit, 1–2 November 2023*. Retrieved March 31, 2025, from https://www.gov.uk/government/publications/ai-safety-summit-2023-the-bletchley-declaration/the-bletchley-declaration-by-countries-attending-the-ai-safety-summit-1-2-november-2023

AlgorithmWatch. (2025, February 6). *Fighting the power deficiency: The AI energy crisis*. Retrieved March 31, 2025, from https://algorithmwatch.org/en/explainer-ai-energy-consumption/

Amnesty International & Access Now. (2018, May 16). *The Toronto declaration: Protecting the right to equality and non-discrimination in machine learning systems*. Presented at RightsCon 2018, Toronto, Canada. Retrieved March 31, 2025, from https://www.torontodeclaration.org/

Asaro, P. M. (2008). How just could a robot war be? In P. Brey, A. Briggle, & K. Waelbers (Eds.), *Current issues in computing and philosophy* (pp. 50–64). IOS Press. https://philpapers.org/rec/ASAHJC

ASME. (n.d.). *ASME standards*. Retrieved March 31, 2025, from https://www.asme.org/codes-standards

Association for Computing Machinery US Public Policy Council. (2017, January 12). *Statement on algorithmic transparency and accountability*. Retrieved from March 31, 2025, from https://www.acm.org/binaries/content/assets/public-policy/2017_usacm_statement_algorithms.pdf

Association for Computing Machinery's Committee on Professional Ethics. (2018). *2018 ACM Code of Ethics and Professional Conduct: Draft 3*. https://ethics.acm.org/2018-code-draft-3/

Australian Human Rights Commission. (2019). *Human rights and technology: Discussion paper*. https://tech.humanrights.gov.au/

BAAI. (2019). *Beijing AI principles*. Retrieved March 31, 2025, from https://ai-ethics-and-governance.institute/beijing-artificial-intelligence-principles/

Baker & McKenzie International. (2023). *Can a global framework regulate AI ethics? Insight Plus*. Retrieved March 31, 2025, from https://insightplus.bakermckenzie.com/bm/investigations-compliance-ethics/international-can-a-global-framework-regulate-ai-ethics

Bradford, A. (2023). *Digital empires: The global battle to regulate technology*. Oxford University Press.

Broeckling, U. (2015). Heldendämmerung? Der Drohnenkrieg und die Zukunft des militärischen Heroismus. *Behemoth: A Journal on Civilisation, 8*(2), 97–107. https://doi.org/10.6094/behemoth.2015.8.2.871

Cadwalladr, C. (2024, August 3). 'A polarisation engine': how social media has created a 'perfect storm' for UK's far-right riots. *The Guardian*. Retrieved March 31, 2025, from https://www.theguardian.com/media/article/2024/aug/03/a-polarisation-engine-how-social-media-has-created-a-perfect-storm-for-uks-far-right-riots

CNIL, European Data Protection Supervisor (EDPS), & Garante per la Protezione dei Dati Personali. (2018, October 23). *Declaration on ethics and data protection in artificial intelligence*. 40th International Conference of Data Protection and Privacy Commissioners (ICDPP), Brussels, Belgium. Retrieved March 31, 2025, from https://www.privacyconference2018.org/system/files/2018-10/20180922_ICDPPC-40th_AI-Declaration_ADOPTED.pdf

Council of Europe. (2018a). *Algorithms and human rights – Study on the human rights dimensions of automated data processing techniques and possible regulatory implications*. Retrieved March 31, 2025, from https://rm.coe.int/algorithms-and-human-rights-en-rev/16807956b5

Council of Europe. (2018b). *European ethical charter on the use of artificial intelligence (AI) in judicial systems and their environment*. Retrieved March 31, 2025, from https://rm.coe.int/ethical-charter-en-for-publication-4-december-2018/16808f699c

Council of Europe. (2018c). *Recommendation of the Committee of Ministers: Guidelines to respect, protect and fulfil the rights of the child in the digital environment*. Retrieved March 31, 2025, from https://edoc.coe.int/en/children-and-the-internet/7921-guidelines-to-respect-protect-and-fulfil-the-rights-of-the-child-in-the-digital-environment-recommendation-cmrec20187-of-the-committee-of-ministers.html

Council of Europe. (2019a). *Committee of Ministers: Declaration on the manipulative capabilities of algorithmic processes.* Retrieved March 31, 2025, from https://search.coe.int/cm/pages/result_details.aspx?ObjectId=090000168092dd4b

Council of Europe. (2019b). *Consultative Committee of the Convention for the Protection of Individuals with Regard to the Processing of Personal Data (Convention 108): Guidelines on artificial intelligence and data protection.* Retrieved March 31, 2025, from https://rm.coe.int/guidelines-on-artificial-intelligence-and-data-protection/168091f9d8

Council of Europe. (2024). *Council of Europe and artificial intelligence.* Retrieved March 31, 2025, from https://www.coe.int/en/web/artificial-intelligence

D 64 Zentrum für Digitalen Fortschritt. (2018). *Grundwerte in der digitalisierten Gesellschaft. Der Einfluss künstlicher Intelligenz auf Freiheit, Gerechtigkeit und Solidarität.* Retrieved March 31, 2025, from https://d-64.org/wp-content/uploads/2018/11/D64-Grundwerte-KI.pdf

Dalai Lama. (2025). *His Holiness the Dalai Lama Supports Human Rights-Based Data-Based Systems (HRBDS) and the Establishment of IDA at the UN.* Retrieved March 31, 2025, from https://idaonline.ch/2025/03/07/his-holiness-the-dalai-lama-supports-human-rights-based-data-based-systems-hrbds-and-the-establishment-of-ida-at-the-un/

Dutton, I. R. (1990). Engineering code of ethics. *IEEE Potentials, 9*, 30–31. https://doi.org/10.1109/45.65865

EDRIi-network. (2014). *The Charter of Digital Rights. A Guide for Policy-Makers.* Retrieved March 31, 2025, from https://edri.org/wp-content/uploads/2014/06/EDRi_DigitalRightsCharter_web.pdf

El Baradei, M., Nwogugu, E., & Rames, J. (1995). International law and nuclear energy: Overview of the legal framework. *IAEA Bulletin, 3*, 16–25.

European Group on Ethics in Science and New Technologies. (2009). *Ethics of information and communication technologies.* Retrieved March 31, 2025, from https://op.europa.eu/en/publication-detail/-/publication/c35a8ab5-a21d-41ff-b654-8cd6d41f6794/language-en/format-PDF/source-77404276

European Group on Ethics in Science and New Technologies. (2014). *Ethics of security and surveillance technologies.* Retrieved March 31, 2025, from https://op.europa.eu/en/publication-detail/-/publication/6f1b3ce0-2810-4926-b185-54fc3225c969

European Group on Ethics in Science and New Technologies. (2015). *The ethical implications of new health technologies and citizen participation.* Retrieved March 31, 2025, from https://op.europa.eu/en/publication-detail/-/publication/e86c21fa-ef2f-11e5-8529-01aa75ed71a1/language-en/format-PDF/source-77404221

European Group on Ethics in Science and New Technologies. (2018a). *Future of work, future of society.* Retrieved March 31, 2025, from https://op.europa.eu/

en/publication-detail/-/publication/9ee4fad5-eef7-11e9-a32c-01aa75ed71a1/language-en/format-PDF/source-314882261

European Group on Ethics in Science and New Technologies. (2018b). *Statement on artificial intelligence, robotics and 'autonomous' systems*. Retrieved March 31, 2025, from https://op.europa.eu/en/publication-detail/-/publication/dfebe62e-4ce9-11e8-be1d-01aa75ed71a1/language-en

European Parliament. (2023). *Artificial intelligence act: Deal on comprehensive rules for trustworthy AI*. Retrieved March 31, 2025, from https://www.europarl.europa.eu/news/en/press-room/20231206IPR15699/artificial-intelligence-act-deal-on-comprehensive-rules-for-trustworthy-ai

European Union. (2018). *Charter of fundamental digital rights of the European Union*. Retrieved March 31, 2025, from https://digitalcharta.eu/

FAT/ML. (2016). *Principles for accountable algorithms and a social impact statement for algorithms*. Retrieved March 31, 2025, from https://www.fatml.org/resources/principles-for-accountable-algorithms

Francesco. (2023, December 8). *Botschaft Seiner Heiligkeit Papst Franziskus zum 57. Welttag des Friedens: Künstliche Intelligenz und Frieden*. Retrieved March 31, 2025, from https://www.vatican.va/content/francesco/de/messages/peace/documents/20231208-messaggio-57giornatamondiale-pace2024.html

Future of Life Institute. (2017). *Asilomar AI principles*. Retrieved March 31, 2025, from https://futureoflife.org/ai-principles/

Grut, C. (2013). The challenge of autonomous lethal robotics to international humanitarian law. *Journal of Conflict and Security Law, 18*(1), 5–23. https://doi.org/10.1093/jcsl/krt002

Guterres, A. (2023, July 18). *Secretary-general's remarks to the Security Council on artificial intelligence*. UN. https://www.un.org/sg/en/content/sg/speeches/2023-07-18/secretary-generals-remarks-the-security-council-artificial-intelligence

Heikkilae, M. (2022, December 12). The viral AI avatar app Lensa undressed me—without my consent. *MIT Technology Review*. Retrieved March 31, 2025, from https://www.technologyreview.com/2022/12/12/1064751/the-viral-ai-avatar-app-lensa-undressed-me-without-my-consent

High-Level Expert Group on Artificial Intelligence [HLEG AI] of the European Commission. (2019). *Ethics guidelines for trustworthy artificial intelligence*. https://ec.europa.eu/digital-single-market/en/high-level-expert-group-artificial-intelligence

Hood, J. (2015). The equilibrium of violence: Accountability in the age of autonomous weapons systems. *Brigham Young University International Law & Management Review, 11*, 12–40. https://papers.ssrn.com/sol3/papers.cfm?abstract_id=2425371

House of Commons of the United Kingdom – Science and Technology Committee. (2018, May 23). *Algorithms in decision-making* (Report No. 4). Retrieved

March 31, 2025, from https://publications.parliament.uk/pa/cm201719/cmselect/cmsctech/351/351.pdf

IAEA. (2011). *The international legal framework for nuclear security*. Retrieved March 31, 2025, from https://www.iaea.org/publications/8565/the-international-legal-framework-for-nuclear-security

IAEA. (2013). *IAEO Basiswissen: Den Beitrag nuklearer Technik zur Gesellschaft maximieren und ihre friedliche Verwendung verifizieren*. IAEA.

IEEE Standards Association. (2024). *The IEEE global initiative on ethics of autonomous and intelligent systems*. Retrieved March 31, 2025, from https://standards.ieee.org/industry-connections/ec/autonomous-systems.html

IGF. (2014). *The Global Multistakeholder for Dialogue on Internet Governance Issues*. Retrieved March 31, 2025, from http://intgovforum.org/cms/2014/IGFBrochure.pdf

IGF. (2019). *Best Practice Forum on Internet of Things, Big Data, Artificial Intelligence*. Retrieved March 31, 2025, from https://www.intgovforum.org/multilingual/filedepot_download/8398/1915

Isaac, M., & Schleifer, T. (2025, January 7). Meta to end fact-checking program in shift ahead of Trump term. *The New York Times*. Retrieved March 31, 2025, from https://www.nytimes.com/2025/01/07/technology/meta-fact-checking-facebook.html

ISO. (2020). *Information technology – Electronic discovery – Part 3: Code of practice for electronic discovery*. Retrieved March 31, 2025, from https://www.iso.org/standard/78648.html

ITIC. (2017). *AI Policy Principles*. Retrieved March 31, 2025, from www.itic.org/resources/AI-Policy-Principles-FullReport2.pdf

ITU. (2024, November 27). *Global Internet use continues to rise but disparities remain, especially in low-income regions*. https://www.itu.int/en/mediacentre/Pages/PR-2024-11-27-facts-and-figures.aspx

JSAI. (2017). *The Japanese Society for Artificial Intelligence ethical guidelines*. Retrieved March 31, 2025, from https://www.ai-gakkai.or.jp/ai-elsi/wp-content/uploads/sites/19/2017/05/JSAI-Ethical-Guidelines-1.pdf

Kahn, L. (2017). Military robots and the likelihood of armed combat. In P. Lin, R. Jenkins, & K. Abney (Eds.), *Robot ethics 2.0: From autonomous cars to artificial intelligence* (pp. 274–287). Oxford University Press. https://scholar.google.com/scholar_lookup?title=Military+Robots+and+the+Likelihood+of+Armed+Combat&author=Kahn,+L.&publication_year=2017&pages=274%E2%80%93287

Kaspersen, A., & Wallach, W. (2023). *Envisioning modalities for AI governance: A response from AIEI to the UN tech envoy*. In *Carnegie Council for Ethics in International Affairs*. Retrieved March 31, 2025, from https://www.carnegiecouncil.org/media/article/envisioning-modalities-ai-governance-tech-envoy

Kaufmann, S. (2010). Der 'digitale Soldat': Eine Figur an der Front der Informationsgesellschaft. In M. Apelt (Ed.), *Forschungsthema: Militär* (pp. 271–294). Springer VS.

Kharpal, A. (2017, November 6). Stephen Hawking says A.I. could be 'worst event in the history of our civilization'. *CNBC*. Retrieved March 31, 2025, from https://www.cnbc.com/2017/11/06/stephen-hawking-ai-could-be-worst-event-in-civilization.html

Kirchschlaeger, P. G. (2013a). Human rights as ethical basis for science. *Journal of Law, Information and Science, 22*(2), 1–17. Retrieved March 31, 2025, from https://www8.austlii.edu.au/cgi-bin/viewdoc/au/journals/JlLawInfoSci/2013/2.html

Kirchschlaeger, P. G. (2013b). Religionsfreiheit – ein Menschenrecht im Konflikt. *Freiburger Zeitschrift für Philosophie und Theologie, 60*(2), 353–374.

Kirchschlaeger, P. G. (2013c). *Wie können Menschenrechte begründet werden? Ein für religiöse und säkulare Menschenrechtskonzeptionen anschlussfähiger Ansatz* (ReligionsRecht im Dialog (Vol. 15)). LIT-Verlag.

Kirchschlaeger, P. G. (2016). *Menschenrechte und Religionen: Nichtstaatliche Akteure und ihr Verhältnis zu den Menschenrechten* (Gesellschaft – Ethik – Religion (Vol. 7)). Ferdinand Schoeningh.

Kirchschlaeger, P. G. (2020). Human dignity and human rights: Fostering and protecting pluralism and particularity. *Interdisciplinary Journal for Religion and Transformation in Contemporary Society, 6*(1), 90–106. https://doi.org/10.30965/23642807-00601006

Kirchschlaeger, P. G. (2021). *Digital Transformation and Ethics. Ethical Considerations on the Robotization and Automation of Society and the Economy and the Use of Artificial Intelligence*. Nomos..

Kirchschlaeger, P. G. (2024a). An International Data-Based Systems Agency IDA: Striving for a peaceful, sustainable, and human rights-based future. *Philosophies, 9*(3), 73. https://doi.org/10.3390/philosophies9030073

Kirchschlaeger, P. G. (2024b). Automatisierte Waffensysteme: Eine ethische Analyse. *Stratos – Militärwissenschaftliche Zeitschrift der Schweizer Armee*, 24–40.

Kirchschlaeger, P. G. (2024c, December 2). Protecting children from anti-social media. *Project Syndicate*. Retrieved March 31, 2025, from https://www.project-syndicate.org/commentary/australia-ban-on-children-using-social-media-should-be-emulated-by-peter-g-kirchschlager-2024-12

Kirchschlaeger, P. G. (2024d). Securing a peaceful, sustainable, and humane future through an International Data-Based Systems Agency (IDA) at the UN. Data & Policy, 6, e78. https://doi.org/10.1017/dap.2024.38

Kirchschlaeger, P. G. (2024e). The need for an International Data-Based Systems Agency (IDA) at the UN: Governing 'AI' globally by keeping the planet sustainably and protecting the weaker from the powerful. *Journal of AI Humanities, 18*, 213–248.

Kirchschlaeger, P. G. (2024f). Digitalisierung und "künstliche Intelligenz" – Maschinen oder Menschen am Steuer? *MA RWP Lectures 5*. Zentrum für Religion, Wirtschaft, Politik. Wie eine wiederkäuende Kuh. Retrieved March 31, 2025, from https://www.unilu.ch/fileadmin/fakultaeten/ksf/institute/zrwp/dok/Broschuere_ZRWP_Band5_web.pdf

Kirchschlaeger, P. G. (2025). Artificial intelligence: An analysis from the rights of the child perspective. *Berkeley Journal of International Law*. Retrieved March 31, 2025, from https://www.berkeleyjournalofinternationallaw.com/post/artificial-intelligence-an-analysis-from-the-rights-of-the-child-perspective

Lensa. (2024). *Lensa AI: Photo & video editor*. Retrieved March 31, 2025, from https://lensa.app/

Leys, N. (2018). Autonomous weapon systems and international crises. *Strategic Studies Quarterly, 12*(1), 48–73. https://www.jstor.org/stable/26333877

Mahari, R., & Pataranutaporn, P. (2024, August 5). *We need to prepare for 'addictive intelligence'*. MIT Technology Review. Retrieved March 31, 2025, from https://www.technologyreview.com/2024/08/05/1095600/we-need-to-prepare-for-addictive-intelligence/

Montag, C. (2023, October 19). Was verändern Smart Phones in unserem Denkapparat? [Oral Presentation]. *ISE-Ringvorlesung 'Ethik der digitalen Transformation. Aktuelle Fragen'*. Universität Luzern.

Montreal Declaration. (2018). *For a responsible development of AI*. Retrieved March 31, 2025, from https://montrealdeclaration-responsibleai.com/the-declaration/

Nevejans, N. (2016). *European civil law rules in robotics: Study for the JURI Committee*. European Parliament. Retrieved March 31, 2025, from https://www.europarl.europa.eu/RegData/etudes/STUD/2016/571379/IPOL_STU(2016)571379_EN.pdf

Nichols, M. (2023, June 12). UN chief backs idea of global AI watchdog like nuclear agency. *Reuters*. Retrieved March 31, 2025, from https://www.reuters.com/technology/un-chief-backs-idea-global-ai-watchdog-like-nuclear-agency-2023-06-12/

Nosthoff, A.-V., & Maschewski, F. (2019). *Die Gesellschaft der Wearables: Digitale Verführung und soziale Kontrolle*. Nicolai Publishing & Intelligence GmbH.

NSPE. (2019). Code of ethics for engineers. Retrieved March 31, 2025, from https://www.nspe.org/resources/ethics/code-ethics

OECD. (2019). *Recommendation of the Council on artificial intelligence. OECD Legal Instruments*. Retrieved March 31, 2025, from https://legalinstruments.oecd.org/api/print?ids=648&lang=en

Pappenberger, M. (2013). Schattenkriege im 21. Jahrhundert: Die Automatisierung des Krieges durch Drohnen und Roboterwaffen. *Forum Pazifismus, 2*, 38–44.

Partnership on AI. (2024). *Tenets*. Retrieved March 31, 2025, from https://partnershiponai.org/about/#tenets

Price, A. (2018). *First international standards committee for entire AI ecosystem. IE e-tech, 3*. Retrieved March 31, 2025, from https://etech.iec.ch/issue/2018-03/first international standards-committee-for-entire-ai-ecosystem

Ren, S., & Wierman, A. (2024, July 15). *The uneven distribution of AI's environmental impacts*. Harvard Business Review. https://hbr.org/2024/07/the-uneven-distribution-of-ais-environmental-impacts

Richtel, M. (2023, October 25). Is social media addictive? Here's what the science says. *The New York Times*. Retrieved March 31, 2025, from https://www.nytimes.com/2023/10/25/health/social-media-addiction.html

Rockwood, L. (2013). *Legal framework for IAEA safeguards*. IAEA.

Roff, H. M. (2014). The strategic robot problem: Lethal autonomous weapons in war. *Journal of Military Ethics, 13*(3), 211–227. https://doi.org/10.1080/15027570.2014.975010

Santelli, F. (2024, January 18). *Sam Altman: 'In pochi anni l'IA sarà inarrestabile, serve un'agenzia come per l'energia atomica'*. La Repubblica. Retrieved March 31, 2025, from https://www.repubblica.it/economia/2024/01/18/news/sam_altman_in_pochi _anni_lia_sara_inarrestabile_serve_unagenzia_come_per_lenergia_atomica-421905376/

Sharkey, N. (2011). The automation and proliferation of military drones and the protection of civilians. *Law, Innovation and Technology, 3*(2), 229–240. https://doi.org/10.5235/175799611798204914

Sharma, S. K. (1995). The IAEA and the UN family: Networks of nuclear co-operation. *IAEA Bulletin, 3*, 10–15.

Singer, P. W. (2009). *Wired for war: The robotics revolution and conflict in the 21st century*. Penguin Press.

Snow, O. (2022, December 7). Magic avatar app Lensa generated nudes from my childhood photos. *WIRED*. https://www.wired.com/story/lensa-artificial-intelligence-csem/

Stoiber, C., Baer, A., Pelzer, N., & Tonhauser, W. (2003). *Handbook on nuclear law*. IAEA.

Suleyman, M., & Bhaskar, M. (2023). *The coming wave: Technology, power, and the twenty-first century's greatest dilemma*. The Bodley Head.

Swiss AI Association. (n.d.). *Swiss AI association*. Retrieved March 31, 2025, from https://www.swiss-ai.org/

Swiss Federal Council. (2020). *Guidelines on artificial intelligence for the federal government: A framework for dealing with artificial intelligence in the federal administration*. Bern, Switzerland.

The Elders. (2023, May 31). *The Elders urge global co-operation to manage risks and share benefits of AI*. The Elders. Retrieved March 31, 2025, from https://theelders.org/news/elders-urge-global-co-operation-manage-risks-and-share-benefits-ai

The Future of Privacy Forum. (2018). *Beyond explainability: A practical guide to managing risk in machine learning models*. Retrieved March 31, 2025, from https://fpf.org/wp-content/uploads/2018/06/Beyond-Explainability.pdf

The Public Voice Coalition. (2018, October 23). *Universal guidelines on artificial intelligence (UGAI)*. Presented in Brussels. Retrieved March 31, 2025, from https://thepublicvoice.org/ai-universal-guidelines/

Tiku, N., & De Vynck, G. (2025, February 4). Google drops pledge not to use AI for weapons or surveillance. *The Washington Post*. Retrieved from https://www.washingtonpost.com/technology/2025/02/04/google-ai-policies-weapons-harm/

Türk, V. (2023, July 12). *Artificial intelligence must be grounded in human rights, says High Commissioner*. Office of the High Commissioner for Human Rights. https://www.ohchr.org/en/statements/2023/07/artificial-intelligence-must-be-grounded-human-rights-says-high-commissioner

UN. (2024). *Summit of the future 2024*. Retrieved March 31, 2025, from https://www.un.org/en/summit-of-the-future

UN & International Labour Organization. (2024). *Mind the AI divide: Shaping a global perspective on the future of work*. Retrieved March 31, 2025, from https://www.ilo.org/publications/major-publications/mind-ai-divide-shaping-global-perspective-future-work

UN Environment Programme. (1987). *Montreal Protocol on substances that deplete the ozone layer*. Retrieved March 31, 2025, from https://ozone.unep.org/treaties/montreal-protocol

UNESCO & COMEST. (2017). Report of COMEST on Robotics Ethics. Retrieved March 31, 2025, from https://unesdoc.unesco.org/ark:/48223/pf0000253952

UN General Assembly. (2024, March 21). *Seizing the opportunities of safe, secure and trustworthy artificial intelligence systems for sustainable development* (A/RES/78/265). https://undocs.org/en/A/RES/78/265opencasebook.org+3

UN Human Rights Council. (2023, July 14). *Resolution 53/29: New and emerging digital technologies and human rights*. https://undocs.org/en/A/HRC/RES/53/29

UNI Global Union. (2017). *Top 10 principles für ethical artificial intelligence*. Retrieved March 31, 2025, from www.thefutureworldofwork.org/media/35420/uni_ethical_ai.pdf

UNICEF. (2017). *The state of the world's children: Children in a digital world*. Retrieved March 31, 2025, from https://www.unicef.lu/site-root/wp-content/uploads/2017/12/SOWC2017_ENG_lores.pdf?_adin=132415900

Verma, P., & Zakrzewski, C. (2024, April 23). AI deepfakes threaten to upend global elections. No one can stop them. *The Washington Post*. Retrieved March 31, 2025, from https://www.washingtonpost.com/technology/2024/04/23/ai-deepfake-election-2024-us-india/

Villani, C. (2018). *For a meaningful artificial intelligence: Towards a French and European strategy* (Report commissioned by the Prime Minister of France, Édouard Philippe). Retrieved March 31, 2025, from https://www.aiforhumanity.fr/pdfs/MissionVillani_Report_ENG-VF.pdf

Wagner, M. (2014). The dehumanization of international humanitarian law: Legal, ethical, and political implications of autonomous weapon systems. *Vanderbilt Journal of Transnational Law, 47*(5), 1371–1410. https://scholarship.law.vanderbilt.edu/vjtl/vol47/iss5/4/

Wallach, W. (2015). *A dangerous master: How to keep technology from slipping beyond our control.* Basic Books.

WeGovNow. (2020). *Towards #WeGovernment: Collective and participative approaches for addressing local policy challenges.* Retrieved March 31, 2025, from https://wegovnow.eu//

Weisbrod, L. (2025, January 23). Shoshana Zuboff: "Ein Gefängnis ohne Fluchtmöglichkeit." *Die Zeit, 4*, 43.

World Economic Forum. (2022, August 25). *How can the aviation industry make AI safer?* Retrieved March 31, 2025, from https://www.weforum.org/agenda/2022/08/how-can-aviation-industry-make-ai-safer/

World Summit on the Information Society. Declaration of principles, building the information society: A global challenge in the new millennium. *World Summit on the Information Society*, Geneva, Switzerland, 12 December 2003. Retrieved March 31, 2025, from https://www.itu.int/net/wsis/docs/geneva/official/dop.html

Zuboff, S. (2019). *The age of surveillance capitalism: The fight for a human future at the new frontier of power.* Public Affairs.

Index

P. G. Kirchschlaeger, *Ethics and the Digital Transformation of Human Work*, https://doi.org/10.1007/978-3-031-98704-5

The manufacturer's authorised representative in the EU is Springer Nature Customer Service Centre GmbH, Europaplatz 3, 69115 Heidelberg, Germany. If you have any concerns regarding our products, please contact ProductSafety@springernature.com

Printed and bound by CPI Group (UK) Ltd, Croydon, CR0 4YY

19/08/2026

02200168-0001